PRACTICAL LANDSCAPING

IN CENTRAL FLORIDA

Rondi Niles

HOW TO DESIGN AND SELECT PLANTS FOR YOUR LANDSCAPE

Practical Landscaping in Central Florida

Copyright © 2025 by Rondi Niles

Publisher's Cataloging-in-Publication data

Names: Niles, Rondi

Title: Practical Landscaping in Central Florida / Rondi Niles.

Description: The Villages, FL: Hallard Press, LLC, 2025.

Identifiers: LCCN: 2024918821|

ISBN: 978-1-962326-42-1 (paperback)

Subjects: LCSH Vegetable gardening – Southern States| Vegetables. | Cooking. | BISAC GARDENING / Landscape | GARDENING / Lawns | GARDENING / Regional / South (AL, AR, FL, GA, KY, LA, MS, NC, SC, TN, VA, WV)

Classification: LCC SB321.5 .N5 2024 | DDC 635.7--dc23

DEDICATION

This book is dedicated to my husband, Ed. Although I was the one to obtain the formal gardening education and work in the industry, he has enjoyed working outside alongside me for years, both up north and here in Florida. He was finally forced to admit how much he knew one day when he was talking to a gardening friend of his and mentioned how great our unusual variety of variegated hostas looked. The surprised look on my face that he actually knew what a "hosta" was, instead of his usual, "that green and white plant over there," was hysterical.

His out-of-the-box thinking has led us to create some beautiful designs on our properties, incorporating picturesque trees, shrubs, flowers, courtyards, waterfalls, ponds, patios, Floridian block walls, and more. My dream partner — he installs the hardscape and digs the holes, while I choose, place, and plant. Love him forever.

TABLE OF CONTENTS

Preface 1
Introduction 3
PART 1 - THE ART OF DESIGN 7
Yard Design 9
    Community Regulations 10
    Landscapers 14
    Design Steps 19
    Creating Beds and Edging 22
    Mulch 26
    Patios 29
    Driveways 31
    Other Features 34
Designing with Plants 39
    Using Color Properly 40
    Plant Design Elements 42
    Privacy 47
    Plant Size and Tricks 48
    Light Conditions 50
    Hardiness Zones 52
    Specialty Gardens 56
    Bringing It All Together 59
Container Design 63
    Containers 64
    Filling Your Container 66
    Container Design 68
PART 2 - IMPLEMENTATION AND CARE 71
Implementation 73
    Florida Soil 74
    Planting Timeframes 77

- Selecting Specimens...79
- How to Plant...82
- Moving Plants...85

Plant & Lawn Care...87
- Watering Principles...88
- Watering Plants...90
- Fertilizing Plants...92
- Pruning Plants...95
- Weeding Beds...101
- Handling Freezes...105
- Plant Pests...107
- Plant Diseases...111
- Palm Care...112
- Lawn Care...115

PART 3 - PLANTS...125
- How This Part Works...126
- Palms...127
- Trees and Large Shrubs...137
- Shrubs, Hardy...151
- Shrubs, Tropical...164
- Perennial Flowers...182
- Annual Flowers...199
- Bulbs...208
- Succulents...209
- Grasses...212
- Ground Covers...216
- Vines...218
- Shade Foliage...223
- Edibles...228

Epilogue...236

PREFACE

I never considered moving to Florida, just too much heat and humidity. Did you ever say that? And then, one day, my husband and I took a vacation to visit family and friends here. It was late fall, and back in New England where we are initially from, the leaves had fallen, the cold was settling in, and that grey look and feel that would last for almost six months was upon us. But as we pulled into the great state of Florida, there was sunshine, warmth, and palm trees, and a happy feeling of "vacation" engulfed us. We started to wonder, "Why wouldn't we want that all the time?" No more raking leaves, no more shoveling snow, and a permanent vitamin D high.

Eventually fate played a hand, and we (semi) moved. Still not convinced that summer in Florida was a trade-off for winter in New England though, we dipped our toes in as snowbirds in 2011. I had a hard time even thinking about leaving our beautiful property up north that we had designed and created, including gardens galore, a courtyard, waterfall and pond, patios, arbors and pergolas, greenhouses, unusual and rare plants, and more. We planned to continue as snowbirds for at least twenty years, but within four, Florida called to us more than New England and we made the move permanent.

I must admit, I had a hard time adjusting horticulturally when I first arrived. Between horticulture school and accredited nursery training, I had to know the common and Latin names, plus be able to identify by sight, over 500 different plants that commonly grew in my New England area. And as a plant-a-holic, I personally grew many of them over the years. In Central Florida, however, it seemed as if there were only 25 that grew successfully.

Because of my professional gardening background and always being a plant enthusiast, I got to work. I learned my plants here by working in garden centers, visiting trade shows and plant fairs, talking to installation and maintenance landscapers, attending every landscape or plant lecture I could find, learning through clubs from others who had come here before me, growing as many different plants on my property as possible, and becoming a Florida master gardener and helping people with their landscape problems.

I eventually adjusted, and now I appreciate all that Florida has to offer, including that heat and humidity (it is indeed better than winter) and, of course, the plants! My northern greenhouses have been traded in for a beautiful Floridian lanai overlooking natural woods, waterfall and pond, and flowering trees and gardens.

My goal now is to spread gardening and landscaping knowledge to those who have come after me. Besides teaching gardening and landscaping classes since 2014 and overseeing the Landscape & Garden Club for my community, this book is another way to accomplish that. I hope you find this book a welcome addition to your Central Florida gardening library.

INTRODUCTION

You probably figured out quickly that Central Florida is very different from where you lived before. In fact, that is probably why you moved here!

In contrast to northern parts of the country, Central Florida is a semi-tropical location. (Even South Florida is different, being tropical.) We do get occasional frosts and freezes, although snow is an extremely rare occurrence. We also have a rainy season, complete with abundant humidity, and a dry season. All these variable factors can make it challenging for plants, as well as their owners who grow them.

Before we delve into how to create and maintain landscapes in this wonderful place, let us start with some basic concepts.

FLORIDA-FRIENDLY LANDSCAPING

In 2009, the Florida Legislature declared that a deed restriction (like specific communities have) may not prohibit any property owner from implementing Florida-friendly landscaping on his/her land.

You may have heard of the term, but do you know what it actually means? It is a landscape that has been designed, installed, and maintained to reduce environmental impact, according to 9 principles. These principles are as follows:

1. Right plant, right place
2. Water efficiency
3. Fertilize appropriately
4. Mulch
5. Attract wildlife
6. Manage yard pests responsibly
7. Recycle
8. Reduce stormwater runoff
9. Protect the waterfront

You should follow these principles because managing your yard responsibly is very important. It is hard to be a gardener and not care about the environment. These principles will be referred to as we progress in this book. But notice these principles do not mention the art of landscape design, something we will cover extensively in this book.

NATIVE PLANTS

So, does using Florida-friendly plants mean you need to use native plants? No, it does not. The definition of Florida-friendly says "...native and non-native plants all can have useful and beneficial places in Florida-friendly landscapes, provided they are planted in the right place and are maintained according to the other eight key principles of Florida-friendly Landscaping..."

Native plantings

There are some advantages to using native plants. They are acclimated to our tough local environment and they may attract pollinators more than non-native plants. Many self-seed and create a cottage-garden-type look, so if that is not your goal, be sure to buy native plants that will match your existing garden style and not require extra maintenance to keep in check. You will most likely need to visit a specialized native nursery to get true native plants.

Native ground cover plants can be used as a lawn or mulch replacement.

Lawn replacement

Mulch replacement

Most plants sold at local Central Florida nurseries and planted in your yard are indeed Florida-friendly and successful for our area.

WHAT THE BOOK COVERS

This book is meant to be practical, which means it is not general "textbook" information but rather useful and applicable to our specific area and community. It will tell you the real story and provide insider secrets for a lower maintenance yard. And it will present plants that can be easily found (whereas existing reference books are far more generalized). And just because you find a plant at a garden center does not necessarily mean it is right for our area.

So, whether you are looking to renovate your entire landscape, add a few plants, or just learn more about landscaping specifically for our local area, this book will provide information on how to proceed. Covered will be tips on Central Floridian differences from other areas of the country, proper plant placement, how to use color, local hardscape options, approval guidelines, best planting timeframes, specialty gardens (such as butterflies, birds, and bees), and more. You will learn about plants that can be used successfully and vital tips on how to care for them.

The book is divided into three parts. The first part will explain landscape design, the second part will cover implementation and plant care, and the third part will cover the plants, including palms, trees, shrubs, perennial and annual flowers, and more.

PART 1 - THE ART OF DESIGN

Part 1 - The Art of Design

YARD DESIGN

COMMUNITY REGULATIONS

Before you even start planning a new landscape or modifying it, you will need to verify if your community has deed restrictions or HOA- (homeowner association) type regulations that must be adhered to, and become familiar with them for your design. The following is how it works in my community.

Changes to an existing landscape must have a plan submitted to and reviewed/approved by an architectural review committee. Deed restrictions govern what the property line setback requirements are, hardscape requirements, statuary usage (it cannot be used except in birdcages and enclosed courtyards), how high hedges can be, etc. Hardscape is defined as solid materials such as stone, concrete, or brick, compared to softer materials such as plants and soil.

House using beautiful statuary

CHANGING REGULATIONS

Regulations can change over time, but if you had your official approval stamp for previous plans, they would be grandfathered. One example of such a change is setbacks. Years ago, plants were allowed to be grown to the property line (if hardscape was not involved, which is a different setback regulation). But more recently, changes were made to adopt a 2-foot setback. I believe at least one of the reasons was because of swales.

Swales are depressions in a lawn, in between homes, used to channel water away from the houses to prevent water damage, especially in our rainy season when we get deluged with rain.

Swale

Part 1 - The Art of Design

When we initially moved here, the house behind us landscaped to their property line and into the swale, effectively removing it. Unfortunately, it was a very wet area, cre-

Swale before, showing drain

Swale after landscaping

ated by the houses on my street being above grade with a slope in the back, and causing mini lakes to form at the bottom of the slope near the houses behind us. The swale had channeled the water to a nearby concrete stormwater catch basin in the lawn between the houses.

By removing the swale, the water could not effectively drain, hence it backed up. And all the plants that the landscaper of the neighbor behind us had installed, where the swale was previously located, died. The catch basin and swale should have provided a clue to the landscaper that this was a wet area even prior to the initial installation, but some landscapers are not that knowledgeable.

Swale removal ramifications

The dead plants were replaced by the homeowner's landscaper several times. Eventually the landscapers created a berm to raise the plants above the water, and then most of the plants lived, but it still did not address the lake problem. If the 2-foot setback had been in place at that time (assuming they had gone through the review process), this situation would not have been allowed, and it would have prevented the drainage problems, which can sometimes lead to sinkholes in Florida.

NOT FOLLOWING THE RULES

Unfortunately, many homeowners do not go through the review/approval process for whatever reason. If a landscaper is hired, they may create a plan and go through the process for you. Whether they do or not, the homeowner is ultimately responsible. Especially when new to the community, homeowners do not always realize there is such a process, and many landscapers do not understand local policies either.

Problems can result from not following the rules. There have been many instances where a homeowner did something inappropriately and without approval. All of a sudden, someone – a new neighbor, a troll driving through the community – will call in a complaint. A deed compliance officer will come out and inspect, and the homeowner will need to rectify the situation. Removal could be expensive. After warnings, the homeowner will eventually be fined and perhaps taken to court.

We also have an anonymous complaint system in my community, which means anyone – even someone from outside the community - can call in a complaint against someone else. This pits neighbor against neighbor, leaving someone to wonder who betrayed them by calling in a minor complaint like beautiful statuary. Another common situation occurs when someone sells their house with non-compliant landscaping, and the new owner, who usually does not know any better, is saddled with a problem if someone complains (like rock mulch being used where it is not allowed).

Properly set back from road

In a different scenario, my neighborhood was subject to a difficult, unhappy neighbor taking advantage of the system. The person called in numerous times but made up the complaints (in other words, nothing was a legitimate problem) just to harass the other neighbors. Many neighborhoods are now thankfully doing away with the anonymous component.

The moral of the story is – follow any local rules to avoid future problems and potentially costly changes.

But what if you have already created your new landscape and only now do you realize the process (which is not a valid excuse, by the way)? I still recommend going through with the approval process, without saying, of course, that you have already implemented your landscape. You will most likely be fine, and at least then you are safe from future problems or regulation changes. If necessary, you can at least correct any problems proactively.

Part 1 - The Art of Design

SUBMITTING A PLAN FOR APPROVAL

A plan does not need to be exceptionally fancy. The plan should show existing house and garden measurements, property lines, and the changes you wish to make.

LANDSCAPERS

QUALIFICATIONS

Landscapers in the state of Florida do not have to be licensed and there is not any industry oversight. Arborists, who work specifically on trees, and landscape architects, who work mostly on large-scale commercial projects, do have to be licensed though. There is a certified landscaper training program in Florida (not a license), but many do not choose to go through it or are unaware of it (and the design portion is limited).

Therefore, anyone can hang a sign on their truck and call themselves landscapers. Particularly in my community, which includes a lot of seniors who no longer wish to do manual labor themselves, new start-up landscapers can easily take advantage of this.

New neighborhoods, where landscaping services are more likely needed than in established neighborhoods, are particularly susceptible. I have seen trucks park in such neighborhoods, hoping that people will go by and think they are working on a neighbor's yard, and therefore if their neighbor hired them, they must be good.

I have also seen the same design mistakes replicated throughout a neighborhood, and I know that they all used the same landscaper just because their neighbor did. Many landscapers leave their company sign in a yard when work is completed, hoping to get more work through their advertising. There is nothing wrong with that, but you truly will not know how good a landscaper is until a yard has gone through at least a full year's cycle of growth and weather.

I find that sometimes landscapers think they know their plants, but many do not know the art of landscape design.

SOLICITING

Soliciting is not allowed in my community, but is unenforceable. Homeowners are instructed to call the sheriff, but by then the solicitor has long since left their property. If a landscaper knocks on your door trying to sell their services, you have to wonder why they need the work. There was a situation where a landscaper had to get business that way because he had swindled so many people. And it is common for "landscapers" to sell palms out of their pick-up truck. Sometimes you will find an ad offering a discount, and I always wonder why they need to advertise that way. Most good, qualified landscapers are already busy and will not need to resort to such soliciting tactics.

HOW TO FIND A GOOD LANDSCAPER

If a landscaper is indeed familiar with the rules of your community and offers to submit a plan for review/approval for you, there is a good chance that they do good work, are not afraid of the approval process, and are therefore qualified.

Also, good landscapers will not ask for money until the job is completed. It not only shows how confident they are in their work, but you can feel more assured that they are not going to scam you. A red flag is when a landscaper takes your money first, even a partial amount down, because scammers may not complete the job or not do so to your satisfaction.

Word of mouth is still the most valuable method of finding someone, but just be sure it is from someone who has had their landscape for more than just a few months without problems. Try to collect multiple recommendations.

Drive around and find examples of a yard that you like. If the homeowner is out in their yard, ask them who they used. Homeowners are usually friendly and will be proud of their landscaping, especially with your positive feedback.

Check reviews online, and verify with the Better Business Bureau online database that a company does not have any outstanding complaints. You can also verify with Florida's SunBiz database (sunbiz.org) that a company is a legitimate business.

Good landscapers may be backed up time-wise, but it shows they are in demand for a reason, so have patience. Two months or more is not unreasonable. Be careful, however, of landscapers who may start a job sooner in order to lock in your contract, but then will not come back for several weeks to complete the work. Ask upfront if they will be working continuously once started.

Smaller job landscapers are exceptionally hard to find. If you are buying a small quantity of plants from an independent garden center, they usually will offer installation services.

PRICE ESTIMATES

I just about fell off my chair when we got an estimate to create a pergola for $10,000. My husband ended up building it himself instead (and yes, we sought approval first) for approximately $1,000 in materials. Another well-known contractor in my community gave us an estimate of $10,000 for another change that was very minor, and I later found out that that was the minimum estimate they always give, regardless of job size. Those experiences left me feeling that contractors, in general, may take advantage of people in my community from a cost perspective.

Get multiple estimates, and make sure you are comparing apples to apples. And why would you get varying estimates?

1. Amount of hardscape

Plants are much cheaper here in Florida than up north, as I have found from working in garden centers in both locations. Perhaps it is because they can be grown in Florida without the massive expense of greenhouse heat in the winter, or perhaps because plants can be grown year-round here instead of only during a smaller window (spring through fall) to make a profit.

However, the biggest cost factor is labor, which comes into play with hardscape. For example, block walls require cutting and fitting, and take more time to install than curbing. So, if one landscape design estimate includes a block wall and the other uses curbing (and there are reasons to use each, which we will cover in a future section), that would result in a cost difference.

Part 1 - The Art of Design

2. Plant size

Do you want your landscape to have an immediate impact? Some landscapers cheapen it out by using very small plants. They will tell you small plants will catch up, and they indeed will, but it will take time, especially for slow-growing plants. The old adage is true: the first-year plants sleep, the second year they creep (roots spread), and the third year they leap (foliage explodes).

Skimpy plants

If you start with larger plants to begin with, you will have a stronger root system, which is more important than foliage initially. Trees usually come in 15-gallon or 30-gallon containers, shrubs in 3-gallon and 7-gallon, and flowers in 1-gallon or 3-gallon.

15, 7, 3, 1 gallon pots

3. Plant guarantee

Make sure your landscaper will give you a warranty on plants. Usually, it is for six months, sometimes a year. During that time, should your plant fail (as long as you cared for it properly), they will replace it. It is especially important if you are having a landscape installed as we approach winter and your design includes tropical plants. If they are not well-established before a freeze, they may not make it. Adding this plant assurance will add to the estimate.

4. Formal computer plan

Formal computer plans have the advantage that you can see a 3-D rendition of what your landscape will look like prior to installation. The cost of the computer software (which can be thousands of dollars) and the time it takes to learn that software (some are complicated) will add to the overhead in the design estimate. However, it helps to eliminate misunderstandings and to ensure you get what you are looking for. Many landscapers will do a 2-D drawing instead.

5. Irrigation

Is revising your irrigation piping part of the plan? If you have an irrigation system, adding or changing garden bed lines should be addressed for proper watering, but many landscapers skip this step.

6. Damage

How is the delivery of materials handled? You will need to pick somewhere for them to be placed. Generally, only 24 hours is allowed if placed on the road. Instead, materials may end up on your grass (potentially killing it or causing compaction) or on your driveway (which may impede cars or scrape driveway paint). How will they access your yard without chopping it up? Dust from cutting hardscape will be all over the place – if near your lanai, do they power-wash afterward? Companies may add a buffer to handle any accidental damage.

7. Experience

Newer landscapers may charge less because they are trying to undercut established businesses in order to get their business off the ground. Do you really want someone who has little experience? Ask how long they have been in business.

8. Under-the-table labor

I have mentioned this item in my classes, and sure enough, several people have approached me afterward and said this happened to either themselves or a neighbor. To keep costs down, a landscaping company owner may pay employees "under the table." Thus, they are not officially employees, and as such, they are not covered by employee compensation plans.

How does that impact you? If they get hurt on your property, it falls under your insurance policy, including any deductibles you may have to cover. Ask a landscaper how they pay their crew. An upstanding landscaper will readily say and be proud of the fact that they are paying their employees appropriately, along with associated insurance.

If price estimates come in higher than you were planning, consider a staged approach to implementation. Perhaps add structures and trees first. Or re-do your front yard first and the back yard later.

Part 1 - The Art of Design

DESIGN STEPS

There are many steps involved in landscape design. I will not bore you with all of them. If you are doing your own landscape design and installation, you will still need to think through the most important aspects. And if you hire a landscaper, it is still good to understand what goes into their process. At a high-level there are four steps.

STEP 1 – BASE DIAGRAM

A base diagram shows where your house and driveway are located, measurements, and property lines. My community keeps these on file for most homes in our Warranty Department, so if you did not get a copy when your home was sold to you, you may be able to get one that way. If unavailable, you can create your own, but you also may need to hire a surveyor for your property.

Base diagram

STEP 2 – SITE INVENTORY & ANALYSIS

This step builds upon the base diagram with other items that exist on your property. Some examples:

Structures
  Existing structures
  Windows/doors
  Roof overhangs
  Existing hardscape
  Utility boxes

Topography
  Grade changes
  Drainage issues
  Sunlight direction
  Swales
  Windy areas

Considerations

Views from home/road

Noise levels

Right of ways/easements

Street lights

Plants

Existing plants

STEP 3 – ACTIVITY/NEEDS DIAGRAM

Why are you looking to re-do or enhance your yard? This step identifies what you are looking to change. On the plan diagram, these items are noted by bubbles (circles) for the areas that will be impacted.

Some of the purposes homeowners choose are below, with the most popular ones in my community being to enhance the beauty in front, and to provide privacy in back (our homes are very close together).

- Privacy
- Curb appeal (beauty/views/color)
- Entertaining/relaxation areas (seating/dining)
- Habitat for birds/butterflies
- Wind block
- Increase home value

Part 1 - The Art of Design

STEP 4 – MASTER PLAN

The final version of your plan takes those generalized bubbles and refines them with the following elements:

- Hardscape
- Structures
- Focal points
- Plants

Completed plan showing hardscape implementation

Completed work

If using a landscaper, they may use a computer-generated plan. Or they may hand draw master plans. There are specific industry guidelines on how to create a formal hand-drawn plan (such as using a specific shape for a type of shrub), but many landscapers do not bother with that.

CREATING BEDS AND EDGING

BED LINES

When creating new beds or modifying existing ones, make your bed outlines curved. Smooth curved beds are more pleasing to the eye, with the added benefit that they look more professional.

Nice curved bed lines

Tight curve

New construction will frequently be cookie-cutter, which includes either straight lines or very tight curves. Neither looks as good, and tight curves are hard to mow as well.

EDGING TYPES

Once you have your beds in place, the next step is to edge them. There are different types of edging commonly used in Central Florida.

- **Trimmer.** No physical barrier, where the edge is maintained by a lawn trimmer or weed whacker tool. If you hire out lawn mowing chores, your mowing contractor will typically include this for you.

Trimmer edged

- **Rubber/plastic.** Thin flexible edging, approximately 4-6" tall, usually comes in rolls. Create a thin trench, add edging, and firm soil against it.

Rubber edging

Part 1 - The Art of Design

- **Metal.** Thin metal edging, usually 4-6" tall and green or brown in color. A bit more expensive than rubber, but also a bit better looking. You can trench or try pounding with a mallet to install.

Metal edging

- **Curbing.** Concrete strip about 6" wide that sits atop your bed. It comes in different colors, although typically just grey and clay tones. Different line patterns may be chosen to be stamped into the concrete. It is installed by curbing specialists who mix the color into the concrete and use a special machine to lay it down. It is generally priced by the linear foot. It will need to be sealed, and if the sun fades the color in time, it may need a paint refresh.

Curbing

- **Block walls.** These are small retaining walls, made of synthetic pre-formed concrete stones, each approximately 12" wide by 4" tall, with a rough front. They are normally stacked 2-3 courses high and finished with 2" tall top caps that are glued down. (Be careful walking on them, however, because the glue may disintegrate over time in our heat, and they can pop off, resulting in injury. Guess how I know this!) If you have a ground-level porch with plantings in front of it, you will need a smaller wall height, perhaps just one course plus a top cap. The stones come in several nice-looking color options to match your overall landscape theme.

Block wall

The stones will need to be cut to fit your bed lines and will take more time than other hardscape options. For that reason, it will be more expensive if done by a landscaper. Some people, including my husband, have been known to do this themselves, but it is a messy job that requires effort.

Retaining walls may occasionally need sealing and power-washing because black mold may form in the rainy season in shady areas or if plants tumble over the walls.

- **Natural stone.** These are real stones, found in Florida, that are either a single course to merely outline a bed or stacked a couple of courses high, similar to block walls.

Natural rock wall

If you are using natural stone, they will not be as tightly fitted as the synthetic block walls. For that reason, you may be more apt to find critters making homes in them.

HOW TO CHOOSE

With so many options, which should you choose? I personally think block walls are the nicest-looking option of all, and the stones come in beautiful colors that can match your house color, driveway, patio, or plant color scheme. However, it is the most expensive option, and if cost is your driving factor, choose one of the others.

If you have a grade change in your yard, you should use a block wall, which can be built to accommodate the slope.

Block wall for slope

Block walls can also be used design-wise to break up a large bed by making a circle or other design in the middle.

Long border design

PATHWAYS IN GARDEN BEDS

Pathways can be used in garden beds, and are typically stepping stones such as flagstone. They can provide functionality, such as getting to a water spigot, or perhaps to another area of the yard. They can help prevent compaction in the rest of the bed, by only walking on the pathway. Sometimes pathways are added just for beauty. Ideally, they should be curved, because that is most pleasing to the eye. Lengthier pathways can lend a sense of intrigue as to where they are taking someone.

Pathway leading to a beautiful waterfall

MULCH

Once you have created your beds, you will need to add mulch. Mulching your beds is Florida-friendly landscaping principle #4. It states that organic mulches break down over time, thereby improving soil structure and providing aeration, which benefits plant roots. It also moderates soil moisture and temperature. And it helps to inhibit weeds, resulting in less work. All these improvements help plants thrive.

TYPES OF MULCH

You should not put landscape fabric down under organic mulch. If you have a lot of weeds in the soil already, you can put down several layers of newspaper underneath the mulch. It will smother the weeds, and break down within a few months.

There are various types of organic mulches used in Central Florida.

- **Pine straw.** Sold cheaply, in bales like straw. It breaks down quickly, so it needs to be reapplyied every 6-12 months.

Pine straw

- **Pine bark.** There are two sizes, regular/large and small. You will have to specifically ask for the latter (which I think looks better) because the default is large. Sometimes I think the large pieces are entire tree limbs! Pine bark will only need to be reapplied every few years.

Pine bark

- **Melaleuca.** We do not have a lot of cedar trees in Florida, unlike New England where I am from, and where such mulch was the norm. So, you most likely will not find natural cedar mulch, but you can sometimes find it dyed in either red or black (which I personally think looks gaudy and unnatural in Florida).

Melaleuca mulch

Melaleuca is a good alternative to cedar since it looks similar. It is made from the invasive Melaleuca tree growing in South Florida, so it is a very environmentally conscious product to use, helping save our native trees. Another benefit is that it is termite-resistant. It is not used by many landscapers because it is not sold by the yard (which landscapers get a discount on), but rather only in bagged form. You can buy the bags yourself or have them delivered, and then either put them down yourself or, for larger jobs, find a landscaper who will do it.

Melaleuca bag

- **Cypress.** You will find cypress mulch for sale, but please do not buy it. By creating mulch from these beautiful trees that take a long time to grow, we are depleting our Floridian natural resources. If consumers do not provide demand for this product, maybe it will stop being made and sold.

ROCK MULCH

Notice that listed above are organic mulches. What is missing from the list? Rock. So, is rock (pebbles) considered Florida-friendly? Much to my initial disbelief, it is. Although it does not break down and sustain the plants, it provides some benefits of mulch, such as moderating water and temperatures and preventing weeds (although not as much as people think).

Rock mulch

Rock pebbles may be too hot. I have personally spoken with maintenance landscapers who have watched plants die because they are surrounded by hot pebbles. Roses are particularly susceptible. However, on the positive side, if you use a lot of tropical plants in your yard, it may help them get through a winter.

Rock is a beautiful option and I actually prefer the Floridian and neater look of it compared to organic mulches. It has the added benefit of lower maintenance since it does not have to be replaced every few years. However, it does not help the soil and plant roots.

Here is a creative mulch, especially meant for a golfing community. This home was located next to a golf course, and the homeowner told me he got one too many golf balls in his yard!

Golf ball mulch

APPLYING MULCH

One of the biggest mistakes I see with garden bed care is how mulch is applied. Even landscapers do not do it correctly. It should be applied away from the trunk/stem of a plant - 6" for smaller shrubs and 12-18" for trees.

"Volcano mulching" is when mulch is piled high against the trunk of a tree, such that it forms a volcano shape. Since mulch holds moisture, too closely applied will allow rot. It also provides a habitat for

Volcano mulch

rodents to chew at the base of a plant.

Mulch should be spread to the canopy/drip line of your plants, in a 2-3" layer.

Part 1 - The Art of Design

PATIOS

PATIO MATERIALS

In northern parts of the country, decks are quite popular. Not so in Florida though, perhaps because of termite threats or the fact that wood can break down quicker in our environment. There are several types of stone-like patios typically found in Central Florida.

- **Stamped concrete.** A concrete patio that has a design and colors stamped into it, typically made to look like a flagstone patio. Because it is solid concrete, it is not water-permeable, so it must be slanted appropriately for runoff.
- **Flagstone/pebbles.** Flagstone is a natural stone alternative, and comes in various colors. In between the flagstones are typically small rocks or sand.
- **Paver patio/sand.** Paver patios are probably the most popular. They are made with synthetic stone pavers that come in a variety of colors to match your landscape or bedding borders.

Pretty paver patio

The cracks between the pavers are filled with sand. More recently, there are products on the market that are polymers, that look and act like a mix of concrete and sand. It will not wash out as easily as sand and will prevent ants from digging in the cracks. It is more expensive than sand, being far longer lasting, a nd your landscaper may not have heard about it. It can be found at big box stores.

Sand-like polymer

PATIO SEALING

Regardless of the material, your patio will need to be sealed occasionally to prevent fading. There are two types of sealants – shiny and matte. Shiny will look like it is perpetually wet, whereas matte will look like it has not even been sealed at all. Once you or your landscaper has selected which type you prefer, continue to use that type forever since there have been reports of a milky sheen if mixed.

Part 1 - The Art of Design

DRIVEWAYS

Driveways are considered part of your landscape and should add to the overall cohesiveness of your yard. In the northern part of the country, driveways are typically made of asphalt, which is more sensitive to UV and heat but can withstand cold freezes and thaws better.

In Florida, driveways are concrete and longer lasting. The benefit of concrete is that there are many options for enhancing your driveway. Although both are considered part of your yard, typically landscapers (or homeowners) handle the landscaping itself, and specialty contractors handle the driveway.

DRIVEWAY DESIGN

Sometimes a house style may have an odd spot leading up to the front door, a patch of soil no larger than 2 feet by 2 feet. I guess home architects and landscapers have truly different skill sets because it is very tough to grow a plant in that small a space.

Odd planting spot

Odd spot plantings

I tried, and I have seen others in my community also who tried, to put a plant in that space and a matching plant on the other side, flanking their garage. Within in no time, the plant in that small space cannot keep up in terms of growth with the other plant, and they look unbalanced. It could be a result of alkaline concrete surrounding the spot from the pathway or from house stucco, or else an irrigation/drainage difference.

Before you do anything with your driveway, you may wish to correct the design by filling that spot with concrete. You can include it as part of a front patio, place a planter on it, or add a fountain.

Concrete monster

Another design consideration is to tone

31

down the driveway. I call it the concrete monster because, typically, in my community, the garages are larger than the home front itself, especially with golf cart garages, and therefore the driveways take up a considerable portion of the front yard. And unlike asphalt, which recedes into the background due to its black color, concrete driveways are white and obnoxiously bright in our Floridian sun.

It is best to tone them down, which also enhances the look of the overall yard aesthetic.

Nice driveway tonedown

DRIVEWAY OPTIONS

There are several ways that concrete driveways can be enhanced. All options will need re-sealing every 2-3 years.

- **Concrete.** Leave as is, but you will get rust stains over time from nearby irrigation. You can get it cleaned and sealed to help prevent stains.

Plain concrete

- **Stamped concrete.** Similar to patios, this look can resemble flagstone or other patterns

Stamped concrete

- **Paint.** This is one of the cheapest options that still provides an aesthetic upgrade. Multiple paint colors can be used to match your house. Particles are added to the paint to make the driveway surface non-slip. Beautiful and unique patterns can be created. Sometimes, a

Painted

mural may be painted in the middle, depicting palm trees or nature, done by an artist who has an alliance with a driveway contractor.

- **Pebbles.** Small pebbles are encased in clear epoxy. It is a beautiful and very natural look. The pebbles come in different colors to match your décor. They can also be used as a floor surface on lanais or around pools.

Pebbles

- **Pavers.** Synthetic paver stones are another beautiful alternative, but probably the most expensive because of the labor involved. The pavers come in multiple colors and can match block walls or patios.

Pavers

The concrete at the end of your driveway will be removed, and whole pavers installed there. The whole pavers will help with the pressure of vehicles continually turning in at that spot. The remainder of the driveway will have half-thickness pavers installed over the concrete. You could opt to have your entire driveway done in whole pavers, but that would be exceptionally expensive.

End of driveway paver installation

Although pavers are one of the prettiest options, most homeowners have had individual pavers crack at some point. Be sure to keep extras left over from the job for future repairs since, in several years' time, you may not find the exact color match.

DRIVEWAY SEALING

All driveway options (except concrete) will need to be sealed every 2-3 years.

OTHER FEATURES

STRUCTURES

Structures can be functional, while at the same time providing uniqueness to a property.

Many structures are made from wood, which is stained for a natural look (tan or brown tones) or to match your house or yard color scheme (blue is popular and pretty). They will need to be re-stained every few years since our intense sun tends to fade them.

Another material is PVC, which is typically found in just white. It will be more expensive than wood but will last longer. It may need to be power-washed occasionally to remove any green or black algae resulting from our intense rains.

- **Trellis.** Typically used as a ladder for climbing vines, they can hide unwanted views, such as blocking out utilities or for privacy from neighbors. There are many different sizes and shapes, and they are easy enough to make yourself if you are handy.

Trellis

- **Arbor.** Arched or square arbors, usually 4 feet in width and depth, make a great entryway to a garden or special area. They add intrigue as to what is on the other side. They are made from wood or ready-made white PVC.

Arbor

- **Pergola.** Pergolas are larger and provide a comfortable sense of enclosure, like a room. They can be made from wood or PVC. Being larger, these are usually made and installed by a landscape contractor.

Wood pergola

PVC pergola

- **Birdcage.** Typically used over pools in Florida, however, they can also be used to enclose gardens and patios, in which case it is called a "serenity birdcage." They usually come in bronze, black, or white metal. There are different hole sizes of the actual screen, varying the amount of light that comes through.

Serenity birdcage

FIREPIT

Unlike in northern parts of the country, in Central Florida we can enjoy outdoor firepits throughout the winter months. It is very relaxing to sit with company around a fire.

Conversational firepit

WATER FEATURES

Water features are not only beautiful, but they can also dampen unwanted noise, such as from nearby traffic. Moving water releases beneficial negative ions into the air.

Fountains, in all types of unique materials and sizes considered works of art, are used in the front yard to welcome guests to your beautiful home.

Welcoming fountain

Front yard statuary fountain

Pondless waterfalls are very natural-looking and attractive, and if outside of a birdcage, they can provide a habitat for beneficial wildlife such as birds and butterflies. Despite the constantly running water (which will keep mosquito breeding at bay), they will

Pondless waterfall

form algae in the heat of our summers. And despite using chemicals that supposedly handle it, we still must powerwash ours every two weeks during the hottest months.

Streams, sometimes ending in a small pond, require room and are most frequently used in serenity birdcages surrounded by beautiful plants. If outdoors and using goldfish or koi, you will attract animals like raccoons, so be sure to cover with heavy-duty chicken wire to keep fish safe.

Anne's stream

STATUARY

Stylish statuary has always been a feature of gardens, used elegantly in famous ancestral homes and botanical gardens. Styles vary from classic antiques to sophisticated modern pieces to lovely Floridian birds. Unfortunately, in my community statuary is not allowed, unless in a birdcage or courtyard. It is

Tanya's statuary

probably because beauty is in the eye of the beholder, and certain pieces may be considered tasteless by others.

Part 1 - The Art of Design

LANDSCAPE LIGHTING

Outdoor lighting can be used to light pathways to your front door or to light up focal features in your yard, such as trees and waterfalls. Although lovely, sometimes I question how much others get to see them since, in Central Florida, daylight sticks around longer anyway, and in my community, life slows down after dark.

Landscape lights can be low-voltage electric or solar path lights or twinkle lights. Solar is good for outlying areas where electricity is not available, but they will need replacing every few years. At that point, your original model may no longer be available for matching. Be sure to place them where they will not light up your bedroom at night (guess how I know).

Night lighting

Part 1 - The Art of Design

DESIGNING WITH PLANTS

USING COLOR PROPERLY

SELECTING A COLOR SCHEME

Let's remember back to our grammar school days, when we learned about the color wheel. Complementary colors are on opposite sides of the wheel. For example, orange and blue are complementary. Analogous colors are next to each other on the wheel. For example, red and orange are analogous.

Color wheel

Complementary orange and blue

Analygous red, orange, yellow

These concepts are used in interior home design and landscape design. When selecting your color palette, use 3-4 colors in total. Anything more appears messy and is not pleasing to the eye.

Select your color scheme in this manner:

1) Select your first color, the predominant one you wish to use in your landscape.

2) Select a second color, complimentary or analogous to the first.

3) Select a third (and even fourth) color analogous to the second color.

Part 1 - The Art of Design

INCORPORATING YOUR COLOR SCHEME

When incorporating your chosen colors into your design, do not just think of flowers. Think about foliage as well. There are some yellow and pretty multi-colored foliage plants (for example, crotons).

Color through foliage

And do not use color literally everywhere. Green is soothing to the eye, and your eye needs to rest as a break from constant color. And green plants tend to be more cold hardy, compared to colorful tropical plants. Green makes an excellent backdrop to color. Privacy hedges are typically tall and green, but color can be added by layering colorful smaller plants in front of a hedge.

Color for hedges

COLOR EFFECTS

In many communities, the house is not far from the road. If you are set back, however, warm colors stand out at a distance and create energy. Cool colors recede into the background and are calming. Larger shapes and flowers can also be seen better when set back from the road.

Color pop from the road

When I first arrived in Florida, I realized that it was difficult to find plants in the color scheme I had used up north, which was blue. The majority of Central Florida plants, interestingly, are red and other warm colors. Even local tropical artwork was primarily red when I was trying to decorate my house.

PLANT DESIGN ELEMENTS

To make a landscape interesting instead of monotonous, different plant-combining principles should be followed. There are many nuances to this topic, but we will stick with the three major concepts.

FORM

Form is the shape of a plant once it reaches maturity. Some plants are nicely rounded, others are spiky or vase-shaped, etc. If every plant in your yard is in the same form, it is extremely monotonous. It can be compounded by landscape maintenance contractors who prune every single plant the same way, in mounds.

Mounds galore

Mounding form

Grass-like form

Vase form

Trees are either multi-trunked or standard, which is a single trunk. Some plants may be either, depending on variety, such as crepe myrtles.

Sometimes a shrub is pruned of lower branches by the nursery grower to make it look like a standard, when in fact, that is not its natural form. In that case, those lower branches will typically regrow, called suckers, and you will have to continually prune to keep them at bay.

Hibiscus grown as a standard (vs. normal bush)

TEXTURE

Plants having a coarse texture have large or distinct leaves. Fine-textured plants have small or lacy leaves and are generally used in masses. Medium-textured plants fall in the middle.

Coarse texture   *Fine texture*

Mix different textures, but try not to use a coarse texture immediately next to a fine texture because it is too jarring. Instead, step down one level at a time.

REPETITION

Repetition means using plants with a common feature repeatedly throughout the garden. It could be plants with similar forms or colors. It ties everything together aesthetically, brings order to the garden, and is particularly helpful in long gardens.

Repetition via color (yellow)

Practical Landscaping

PLANT PLACEMENT

PLANTING EN MASSE

There is a concept in the design world called the rule of odd numbers (3, 5, 7). Odd numbers are pleasing to the eye. Plants should be placed in swaths or masses in odd numbers.

Massed plantings

If a single plant is used, it becomes a focal point and is generally reserved for trees or larger shrubs. Depending on the size of your yard, you should not have too many focal points, and they should be tied together for repetition. One common example is using multiple palm trees in a design. They may be differing varieties, but they still have the same general look and feel and tie together nicely.

Palm placement

Another common placement technique is to surround a focal point tree with smaller shrubs or perennial flowers.

Crotons surrounding a large palm

Focal points do not always have to be plants, either. They may be your colored front door or statuary, for example.

Plants can be placed in a shaped mass, typically a circle, or linear grouping, such as along a driveway or in front of a porch.

Even numbers create a more formal presence. Historical mansions tend to have symmetrical gardens. One way even numbers are used successfully in an average home is when two plants are flanking a front entry or garage.

Shrubs along porch

44

Part 1 - The Art of Design

FRAMING THE HOUSE

Your house should be a major feature of your property, and plants can enhance it, as in the following examples.

- The corner of a house is an ideal location for larger shrubs because it helps soften the hard lines of house corners.
- Make a welcoming front door entry area.
- Enhance walkways by using low plants without thorns or attracting bees.
- Consider fragrant plants near walkways or seating areas.
- Foundation plants are those placed in a garden against the front of the house, between the front door and the corners. They generally need to be smaller plants.

Left corner of house softened by small palm

Welcoming entry

BLOCKING HOME FEATURES

One of the biggest mistakes is placing plants where they hide a home's features. In my community, there are house styles with charming bay windows or lovely Floridian arches. Landscapers untrained in the art of design typically will use inappropriately sized plants that will eventually cover these beautiful architectural features.

Blocking arched window

Hiding the entry with a large plant makes it appear very unwelcoming. Also, try not to block windows. Use naturally smaller plants underneath them (or keep continually pruning them below the sill line). If you are purposely using plants to cover your windows for privacy purposes, try to place them further out from the window.

Blocking entry

Keep all plants away from directly touching the house. It will allow better air circulation, prevent mold and mildew, and prevent critters such as rats from reaching your roofline. Another big mistake I see is planting a small tree too close to the house – eventually, it grows up!

Palm planted too close to house

Part 1 - The Art of Design

PRIVACY

Privacy is especially important in my community since houses are very close together.

Privacy around a lanai is particularly desirable in order to block the view of a neighbor's lanai. The best way is to create a patio or walkway near the house, then add plants away from the house and closer to the property line. This provides a view with more depth.

Diane's privacy from inside

Multiple plants provide privacy (ex., pergola privacy)

Use different plants and differing heights. Multi-trunked palms work well for those wanting a tropical look. Add multiple layers, with smaller plants in the front.

Hedges are one of the most popular ways to increase privacy, especially when fences are not allowed.

Privacy hedge

Berm

Another way to increase privacy is by increasing height via berms. Berms are mounds of soil created a foot or two above ground level then planted as normal. They are also used in wet areas to help with drainage.

A courtyard also provides privacy, and plants can be incorporated to provide a beautiful tropical oasis.

Nan's courtyard

PLANT SIZE AND TRICKS

FLORIDIAN SIZE

Florida plants can get BIG! They look small in their little containers at the garden center, but please read the tags for the ultimate size.

Some independent garden centers do not provide tags. Many such garden centers in my local area started out as landscapers who needed somewhere to keep their plants, and since they are always pulling from their stock for installations, they do not bother with tags (tags cost them extra). Have you hired a landscaper and ever received the tags after installation? Most likely not, so ask them about your plants instead.

Big box stores have an advantage in that they always have tags, probably because they do not have enough skilled horticulture personnel to answer questions, and tags provide that information. Tags are also a marketing advantage with their photos, especially when something is not in bloom and customers can still see the bloom on the tag.

Sometimes tags are not provided for devious reasons. While working at a local garden center, customers would ask me, "How big does this plant get?" I was instructed by my manager to ask in return, "How big do you want it to get?" If the customer said 3 feet, but I knew the plant would normally reach 6 feet, I was to tell them, "Yep, you can keep that plant trimmed to 3 feet." If they persisted and said, "But I do not want that maintenance though," I was allowed to indicate whether it was a small, medium, or large plant but not to divulge the actual height. I suspect it was because if customers knew the actual size, they may not buy the plant. (Please note, I did not feel comfortable with that approach.)

Sometimes a plant will have different varieties, each being a different size. In the example below, we purchased our home with a loropetalum, which is normally a bush when used in landscapes. Letting it remain natural, it quickly grew into a tree form instead. We did not know how poorly it looked until we saw the sale photos online! In all fairness, however, we never saw it from the front of the house since we were on a cul-de-sac and only came in from the side

Before pic *After pic* *Normal viewpoint*

direction.

The point is to make sure you are placing a plant where it has room to grow to its full potential. If not, you will end up with excessive pruning maintenance or getting diseases from planting too close.

RESTRICTING SIZE

Nursery growers, who sell to the big box stores and garden centers, will sometimes apply growth hormones. This is not done deviously but rather to help the retail business by keeping the plants smaller longer, thereby increasing shelf life. Otherwise, a business may have to prune to keep the plant around until it is sold. Once the plant is put in the ground, the hormone will wear off. So again, pay attention to the tag for the ultimate plant size.

This become that!

LIGHT CONDITIONS

Florida-friendly landscape principle #1 is called "right plant, right place." It means that plants should be environmentally sited properly based on the location or plant's hardiness zone, water amounts, light requirements, soil conditions, wind tolerance, etc. Doing so reduces irrigation and fertilization needs providing more resistance to pests. In this section, we will discuss light.

SUN

Full sun is considered to be 6 or more hours of sun per day. When you buy a plant in a garden center, it should have a tag indicating the plant's light requirements. Make sure you pay attention to this because in Central Florida, being closer to the equator than much of the rest of the country, the sun is more potent.

Some plants prefer full sun but may "tolerate" shade. But when planted in shade, the flower quantity will be reduced and it will be slower growing. Other plants can have multiple light conditions but really do have a preference. For example, dwarf podocarpus grows 3-5'. We planted it in our part-shade backyard, and it has stayed 2' tall. But in our sunny front yard, it grew to the full 5' tall, and we need to keep it pruned.

Keep in mind that light may change over time, such as larger plants growing up and eventually shading or removing shade from nearby smaller plants. Sometimes a larger plant may become diseased and need removal, and the previously shaded plants near it may suffer in the sudden full sun.

SHADE

There is very little natural shade in my community, in newer sections especially. Sometimes areas are touted to be "natural," implying woods and trees, and perhaps around the perimeter of the entire area. But when a builder creates homesites, the land is usually stripped of trees to get heavy equipment in, lay water lines, and build.

Land cleared for development

Part 1 - The Art of Design

As my community continues to expand and spread south, more houses are being built per acre. Therefore, there is less yard space per home and houses closer to the road, leaving little room for big trees that would provide shade.

New communities with little shade

Tree surrounded by shade plants

It is interesting to see properties in northern sections of my community, where beautiful large oaks can indeed be found.

If you want plants that take shade, try planting them on the north side of your house. You can also try using structures to provide shade, such as a trellis.

Trellis shade

Practical Landscaping

HARDINESS ZONES

ZONE DESIGNATIONS

The USGA puts out a map for the United States, approximately every decade, that shows temperature averages. They are divided into zones called hardiness zones, which indicate how much cold a plant can withstand. You can find out which hardiness zone you are located in by using the zone finder at the USGA website (search online for "USGA plant hardiness map"). With the last map update in November 2023, my entire community is now in Zone 9B, reflecting a warming trend (previously, northern parts were in Zone 9A).

Hardiness map for my community – Zone 9B

There are also heat zones, which indicate how much heat stress a plant can take. Interestingly, Florida and southern California have the same hardiness zones. But if you have been out west, you know the climate is quite different. So, what is missing? A humidity factor!

Plant tags will indicate hardiness zone for a given plant, and the industry is trying to move towards showing heat zone also. But you will not find humidity/dryness. Local growers tend to know what works best in that regard. Select plants for your yard that match your correct zones.

Plant tag with hardiness zone

Part 1 - The Art of Design

HARDY PLANTS

Hardy plants will survive a freeze without damage. If you are looking for low-maintenance hardy plants, pay attention to medians that have been planted by professional landscapers under contract, which means they need to keep labor costs low via low maintenance. They are not going to spend time pruning and caring for plants excessively, and they will not use plants that need to be coddled for freezes.

Hardy green common area

ROOT HARDY

Some plants are referred to as "root hardy." It means top growth may die back during a freeze, but the root system will still be okay, and the plant will grow back in spring. These are usually tropical plants, and they will need help during freezes if you want to prevent the dieback. So why use them? Because they are more colorful than hardy plants, which tend to be primarily green.

Root hardy Croton returning in spring

Cold damage

In my community, many, many tropical plants are used because everyone wants color. If using tropical plants, be sure to include truly hardy plants in your design as well, so you do not end up with an entire yard of dead looking or browned-out plants in winter.

The chances of a tropical plant getting through a freeze can be increased if used as follows:

- The longer a plant has been established, the better it will do.
- Favorable "microclimates" (ex. against the house wall) can improve odds.
- A susceptible plant surrounded by hardier plants will do better.

Ti plants in a microclimate against the house

You may think stores always carry appropriate plants, but that is not true. We loved our red Japanese Maples back in New England, but they do not grow well in Central Florida. We knew that but saw one at a big box store and, missing them, decided to experiment, even if it meant just enjoying it for one season. We planted in an area with a cooler microclimate, being lower than the rest of our yard. It has been five years now and it is still alive, but not as healthy looking as the ones we grew up north.

And spring bulbs? Big box stores have an amazingly beautiful selection of boxed bulbs, such as tulips. Alas, almost all of them will not grow here. My theory is that a southeast regional manager places the order for all their stores in the southeast, and does not realize or does not accommodate the fact that Central Florida is very different.

Independent garden centers have an advantage over big box stores in this regard since they usually carry appropriate stock.

Part 1 - The Art of Design

EVERGREEN VERSUS DECIDUOUS

We do have a fall here in Central Florida, although we get little colorful foliage compared to northern parts of the country. Missing our gorgeous New England sugar maples, we planted a Floridian variety in our yard that comes close.

Fall color on Floridian maple

Deciduous Crepe Myrtle

Although most plants here are evergreen, some are deciduous, meaning they are hardy, but will drop their leaves. Crepe Myrtles are a good example.

To help regulate temperatures in your home and improve heating/cooling costs, plant deciduous trees on the south or west side of your house. They will provide shade in the summer, but in winter after leaves have dropped, they will let sunlight through to warm your house.

If using deciduous trees in your design plan, make sure they are combined with evergreens so your yard does not look totally dead in winter. And if you are a snowbird (meaning you flee north for the summer), be careful about using too many plants that will lose their leaves when you are here in winter or that only flower when you are not here in summer.

Some plants have leaves or needles that turn yellow and sporadically drop in spring. Many people think these plants are dying. What is really happening is that the spring growth of oaks and pines, for example, will push out the old growth before putting out new growth. If you have seasonal spring allergies, when you see this occur, start taking your over-the-counter allergy medicine because pollen is not far behind!

Oak shedding in spring

SPECIALTY GARDENS

Attracting wildlife and pollinators to your yard is one of the Florida-friendly landscaping principles; principle #5. Incorporate these gardens into your design to attract birds with berry bushes, bees, and butterflies with pollinator plants, and reduce the need for insecticides that can harm the 'beneficials' by using plants that are not prone to problems.

Here are some general tips for attracting wildlife.

- Do not use pesticides! These will kill the good bugs along with the bad.
- Plant in masses so wildlife can better find plants.
- Use different plant colors. Different pollinators are attracted to different colors.
- Vary bloom times, so there is always something available for food.
- Provide some shade cover, not just sun.
- Provide a nearby water source.

BEES

Interestingly, honey bees are not native. However, there are several species of native bees in Florida. One of the prettiest is the green sweat bee. Most bury singly in the ground or in logs, and they are not aggressive since they do not have a hive to defend.

Each bee species has its own proboscis (tongue) length for reaching nectar and pollen. However, most are generalists when it comes to plant preference, meaning they do not require specific plants. Many enjoy daisy-like flowers for a landing pad. Bees cannot see red; they prefer white, yellow, blue, and purple flowers.

Bee heading toward Chaste tree

BIRDS

Some birds eat the seeds of flowers and berries from trees and shrubs. Others eat worms and insects. Hummingbirds use nectar plants.

Connie's Bluebird

Connie's Purple Martin family and nesting house

Native plants attract more native insects, which our native insect-feeding birds prefer. Try to attract birds naturally and avoid bird feeders, because the seeds will attract rats, and rats attract snakes.

Since we had many bird feeders up north and they brought us such joy, my husband really wanted a feeder in Florida. We thought if we placed it in our woods, it would be okay. However, we had an outdoor waterfall nearby, and while it was turned off for maintenance, rats decided it was an excellent home – both food and water just for them! They ate through the liner to create their nests, and we had to take the entire waterfall apart to put a new liner in, not an easy task with so much rock. And even though it was a squirrel-proof feeder, the raccoons had a good go at it too (however, my husband won that battle, being the master of rope and twist-ties). We can now say it is a true story.

BUTTERFLIES

Different butterfly species are attracted to different colors and plants. Butterflies have a fascinating life cycle: egg, caterpillar, chrysalis, butterfly.

Caterpillars need host plants to devour, and most host plants are native plants. Place them in the back of your garden, or hide them amongst other plants, because they will be devoured and left looking like sticks until they recover. Interestingly, caterpillars may be different colors based on what they have been eating, for example, green foliage or yellow flowers.

Butterflies use nectar plants, a much wider selection of plants, to get their food. But - they only lay eggs in host plants.

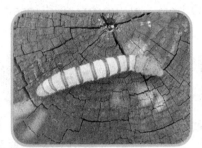

Sulfur butterfly (courtesy of Alycyn from The Villages' Butterfly Club)

Beautiful sulfur caterpillar near yellow Cassia tree

PETS

If you own pets, be sure to take that into consideration, especially if using plants in a birdcage or on a lanai where pets are free to roam. Reference pet-friendly plant lists if need be – aspca.org is generally recognized to be the expert. Oleander, with berries, and Allamanda are extremely dangerous, for example.

Pretty but poisonous Allamanda

I have found that my cats and dogs tend to like grass-like plants to munch on. Cordyline is another poisonous plant that I stay away from because of its grass-like form. Instead, I grow special containers of oat grass and catnip that my beloved pets know are solely for them.

Part 1 - The Art of Design

BRINGING IT ALL TOGETHER

NEW LANDSCAPE CHANGES

Below is how your landscape will typically change with a new plan.
- Concrete driveway -> Painted or pavers
- Square tight garden beds -> Expanded beds with nice curves
- Trimmer edged beds -> Block walls or curbing
- Pine straw mulch -> Stone or pine bark
- Green foundation plantings -> Colorful tropical plants
- Focal point tree (magnolia, holly, crepe myrtle) -> Palms

SELECTING PLANTS

After figuring out your overall yard design, including garden bed placement and structures, it can be overwhelming trying to figure out which plants you should use and how to arrange them. The way design plans are done professionally is to initially designate what type of plants (trees, shrubs, etc.) will be used, their attributes (small, yellow, etc.), and their arrangement, and then add in specific plant varieties that fit that criterion.

Let's go through the steps.

1) Pick a color scheme.

2) Next add palms and trees (if blooming trees, match your color scheme). You should not have too many of these. Be sure they are not placed too close to the house or blocking house features.

3) Shrubs are the backbone of any plan and make up the majority of it. Add shrub locations and arrangements. For example, I may want something taller to soften the corner of the house. In this space, I want 5 small bushes en masse. In front I want a shorter border plant. In that space, I want something to line my driveway. Under the tree, I want a part-shade group of shrubs encircling the tree trunk. Along the property line, I want a privacy hedge. And so forth.

4) Then read through Part 2 of this book and pick out shrubs that will work for each area. Choose correctly for sun or shade. Pay attention to size. Think about foliage/flower color and adhere to your scheme. Leave some green. It is okay to use the same plants in multiple places for repetition.

5) Finish with where you will place perennial flowers or containers.

And please - do not use fake plants. They fade over time, leaves will fall off and blow into neighbors' yards, and most of the time, they are very unrealistic looking.

Easy to spot the fake?

DESIGN IMPORTANCE

When I do onsite landscape consultations, I am first concerned about the proper design. And then I give several suggestions for actual plants that will fit that design.

Sometimes I find that the original design for a property was satisfactory, but the chosen plants were incorrect (for example, tall plants blocking windows). If you have the right base, it is much easier to exchange plants than dig new plant holes or move plants around.

Examples of well-designed homes are below.

EXISTING LANDSCAPES

When you have an existing landscape that you are adding to or changing (rather than starting a new yard plan from scratch), it may be more difficult to incorporate the proper design principles into what is already there, but do the best you can.

For instance, if you are changing out a few plants, determine the primary existing color scheme, and work with that. Notice what design elements are already in place as far as plant forms (mounding, upright) and add new forms or textures that add interest as needed. Do not feel that you need to add new varieties of plants. Adding more of the plants already there enhances repetition.

If you have an existing tree, it will be hard to add new plants underneath due to the established root system, so you may wish to leave that area as is.

Part 1 - The Art of Design

CONTAINER DESIGN

Practical Landscaping

CONTAINERS

Annual plants and flowers, which need to be replaced every season, are typically not grown in the gardens in the front yard of a home. The reasons are practical – every time the plants get replaced seasonally, the soil is disturbed, and therefore, the fine roots of existing nearby plants are impacted. It also brings weed seeds to the surface.

"CONTAINER" OPTIONS

There are several solutions for accommodating annual plants.

- Create a **separate garden bed** as a play area just for annuals, usually in front of perennial plant beds.

Separate garden area ready for planting annuals

- Use an **in-ground pot technique,** whereby a pot is permanently buried in the yard, and annuals are planted in the pot. When removed seasonally, there is no damage to nearby plants since only the plant and soil inside the pot are removed. Pots do not have to be fancy since they will be underground, so simple black grower pots are fine. Usually, 3-gallon pots will suffice, but just make sure the pot has a drainage hole.

In-ground pot technique as shown in a container

- Use **containers or planters**, above ground. They can be placed flanking your garage door (although southern exposure can be very hot on containers), near your front entry, or even in the middle of a garden. They add an artistic design statement to your yard. Containers also dress up a lanai or birdcage.

Raised planter (and solution for a tight spot)

Part 1 - The Art of Design

CONTAINER SIZES

Container sizes will vary. Generally, a container approximately 14-16" will accommodate 3-5 flowers. If you wish to plant each one in its own container, many will be fine in a 10 or 12-inch container (it will vary depending on the plant).

If you put any plant in a container, the size of the container will impact the size of the plant. By constraining the roots, a larger plant will remain smaller.

CONTAINER TYPES

Containers are made from many different materials. I have tried some that have broken apart within a couple of years in our hot, rainy weather, but the best ones for outdoors are as follows.

- **Ceramics** - Very popular because they are artistic looking and come in beautiful colors. They also can be expensive, however.

Colorful Talavera style pottery

- **Plastic** - Long-lasting, weather-resistant, lightweight, and generally cheaper. Plastic containers come in various shapes, including round, square, and urns, with molded forms, and many colors. And if they do not have a drainage hole, one can be easily drilled.

- **Self-watering containers** - May be used, depending on what kind you buy and where you are placing them. They have a reservoir full of water on the bottom that is wicked up into the soil as it dries out, keeping the soil consistently moist. The reservoir will need to be filled every so often, but if your containers are not under irrigation, it cuts down on watering chores overall and helps if you will be away from home for a bit.

Self-watering containers work especially well for plants kept under cover of a front entry, or outdoors in our dry season. However, they do not allow for drainage, and particularly if used outdoors during our humid and rainy season in Florida, it may keep the soil too wet. In that case, be sure to get a model that has an overflow hole so excess moisture can be released. And be sure to keep the fill/overflow hole unclogged from debris and watch for mosquitoes entering. Self-watering containers tend to be more expensive.

FILLING YOUR CONTAINER

Covered drainage hole

Cover the drainage hole of your container with landscape fabric, which lasts a long time, or a coffee filter, which breaks down quicker, but by then, the mix is usually settled anyway. Fill your container initially to 1" below the container rim. After planting, you may need to add additional mix.

Use a mix specifically formulated as either a "potting" or "container" mix. These are soil-less mixtures that provide good drainage. You can also make your own using the same ingredients.

TYPICAL CONTAINER COMPONENTS

- **Peat moss (or coir)** - Peat moss is not a sustainable product. Sustainable means not having a negative impact on the environment, and peat moss can deplete aged bogs. Therefore, the horticulture industry is moving towards coir, which is made from coconut husks. Both peat and coir hold moisture, but coir holds more.
- **Compost** - Compost is organic matter known for its ability to improve soil structure (providing beneficial air pockets for roots) and to hold water and nutrients. There are many types of compost, typically broken down into these categories.
  - **Manures.** These contain mostly nitrogen with some micronutrients.
  - **Leaf litter/mushroom.** These have beneficial microbes and bacteria.
  - **Earthworm castings.** These also have beneficial microbes and bacteria.

Part 1 - The Art of Design

- **Perlite and/or coarse sand** - Perlite is a volcanic product that is heated until it puffs up and is the little white pieces you see in mixes. It provides excellent drainage, something particularly important for vegetables. Sand will also provide excellent drainage. However, use coarse sand and not play sand, which can condense to a cement-like texture.

Perlite

- **Mycorrhizae fungi** - This is a beneficial fungus that attaches to the roots of plants to form a symbiotic relationship. It allows more nutrients to be absorbed by the plant. Another benefit is that it helps plants better tolerate stress, such as transplanting, and diseases.
- **Wetting agent** - A wetting agent makes it easier for the initial wetting of the mix. Otherwise, you may have to soak the soil before it will absorb water.

There are also "complete" products on the market that come with moisture control (and typically fertilizer). Moisture control crystals can also be bought separately. The crystals look like pieces of rock salt when dry. As they absorb moisture, they expand into a mini jellyfish appearance. As the soil dries out, the crystals release the moisture and return back to a rock salt appearance until you water again. Moisture control crystals can hold onto too much moisture, which can be the death of plants in the rainy season. If using, just be absolutely sure you have a drainage hole in your container. I have used them successfully in clay containers, since clay wicks water away from plants.

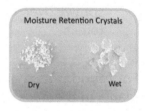

As containers continually get watered, the air pockets that plant roots need will become compressed. For best results, change your soil mix every few years. To stretch it further, you may add fresh compost annually.

CONTAINER DESIGN

PLANT PLACEMENT: THRILLERS, FILLERS, AND SPILLERS

If using a single container for multiple plants, you may have heard of the popular concept—thriller, filler, and spiller.

- **Thriller** – A single taller plant that will make a statement in a container.
- **Filler** – Medium-sized plants, usually rounded in form.
- **Spiller** - Low-growing plants that sprawl over the edge of a container.

For round pots that will be viewed from only one side, place taller plants in the back, filler plants on the sides and middle, and spiller plants in the front. If you plan to rotate your container so it will be viewed from all directions over time, place a taller plant in the center of the container, surrounded by filler plants, with spiller plants in front.

You can bring an interesting design flair to your containers using the following techniques:

- **Add different textures** with grasses or grass-like plants for an architectural statement. Grasses can take over a container with their dense root systems, but I like to use yucca plants or taller flowers.
- **Add different colored foliage** for interest, for example, variegated plants or red/purple foliage plants.
- **Place thrillers and spillers off-center,** which creates a more artistic look.
- Although containers are used functionally for annual plants, you can also **use perennial plants, flowers, or grasses,** which add even more plant combination options.

Part 1 - The Art of Design

Some nicely designed containers are shown below.

GROUPING MULTIPLE POTS

The guidelines above apply to a single container with multiple plants. Another option is to use several containers grouped together, each with a single plant. Use different-sized pots but with a cohesive theme, such as in the same color family. This makes a beautiful design statement.

PART 2 - IMPLEMENTATION AND CARE

IMPLEMENTATION

Practical Landscaping

FLORIDA SOIL

IS THAT REALLY SOIL?

Yes indeed, Central Florida "soil" is mostly sand (except near lakes, where it may be more clay). The good news is that sand provides excellent drainage! On average, this is our soil composition:

- **Clay** (10%)
- **Silt** (30%)
- **Sand** (60%)

Soil layers from new homesite

If your landscaper is creating a new garden bed for you, they will use what is called a wholesale "Florida mix," composed of sand, peat, and compost. If you are a do-it-yourselfer, you will need to get bagged "garden soil" (not a potting mix). It, too, will be

Florida mix

formulated for Florida. I was amazed the first time I bought a bag. I thought something was very wrong with it because it looked terrible compared to what I was used to buying back in New England. I even bought a second bag, from a different brand, thinking just the first one was bad. Come to find out, they do regionalize the mixes. Of course, that tells you about the quality of our soil here. But Florida-friendly plants are acclimated to it.

In New England, we had earthworms everywhere in our soil. They are very beneficial and indicative of healthy soil. Once in Florida, however, I would never see any, instead just the occa-

An actual earthworm in my Floridian yard

sional millipede. And then one time, after a hurricane brought significant rainfall, I was pleasantly surprised to find that I do indeed have worms in my yard! They must have been flushed to the top by all the rain. Our efforts of adding compost to our gardens and lawns, as well as using organic mulch, had paid off.

COMPOST

If you are re-doing a bed, you can add compost. It is also good to add to lawns. Compost improves soil structure, which adds air for healthy roots. It also improves the ability to hold water and nutrients. Because sand is porous and because of our heavy rains in Central Florida, compost does not last forever.

The biggest mistake non-gardeners make is adding compost to just the planting hole. What happens is the plant thinks to itself, "Why do I want to extend my roots further into the soil when it is pretty darn nice here in this compost hole." As a result, your plant will not develop a strong root system or be as healthy.

SOIL PH

Trucks dumping concrete on an end lot

If you are having difficulty with plants in your yard, do you have an end or corner lot? We watched a new neighborhood being built in my community, and during the construction process, excess concrete from cement trucks was repeatedly dumped on an easy-to-access corner lot in that neighborhood. It accumulated there until one day a contractor brought a skid steer to load the concrete into a dump truck for removal. Do you really think they were able to remove all the small pieces? Shortly afterward, they started construction on that lot. We also saw, during the building of our own home here, contractors leaving extra concrete from pouring our steps right on the ground, which they eventually just covered with topsoil.

The problem is three-fold. First, heavy concrete piles will compact your soil. This can also happen, by the way, when you have work done in your yard post-sale, such as putting in a pool, block walls, etc. Secondly, concrete displaces soil, which plants obviously need to thrive. And thirdly, concrete (and stucco) is alkaline. If you are having difficulty with acidic plants, this could be why.

The average soil pH for Central Florida is slightly acidic, good for most plants. But if your pH is too low (acidic), nutrients do not get absorbed, including from fertilizers, no matter how much you put down. And beneficial microorganisms cannot thrive.

If you are having problems, you can get your soil tested for pH and nutrient levels through the University of Florida's soil testing lab, and you will be given remedy advice. Forms and instructions are available online or through your local County Extension Service. Although it is difficult to change the pH long-term, especially if too alkaline, you can try adding compost.

Part 2 - Implementation and Care

PLANTING TIMEFRAMES

FLORIDA SEASONS

Here in Central Florida, we have two seasons:
- **Warm** season (May through October).
- **Cool** season (November through April).

Plants that stay in the ground year after year, such as trees, shrubs, etc., are perennial plants. Seasonal aspects come into play with annual plants and flowers, which only last one season.

In our semi-tropical environment, we also have two other seasons:
- **Rainy** season (June through fall).
- **Dry** season (winter to spring, with a particularly dry spell in April and May).

WHEN TO PLANT

Landscapers and garden centers, who want sales year-round of course, will tell you that containerized plants can be planted any time of the year. Is that a sales pitch? No, it is true, and with proper care, you should be fine. However, there are some times that are better than others.

The rainy season is good because you will get supplemental watering by nature. (However, rain alone probably will not be enough for a new plant initially.) The dry season is harder because plants especially rely on you to keep them adequately watered.

Fall is a good time to plant because root systems are still expanding while the soil is warm, without expending energy growing foliage. Winter could have ramifications, especially with tropical plants, if we get a freeze because new plants are not fully established yet. In that case, be sure to cover your plants (see Cold Hardiness section) or ensure you have a plant warranty from your landscaper.

In the heat of summer, planting in the morning can help new plants adjust to heat stress.

STOCK AVAILABILITY

In northern parts of the country, a huge plant stock flush happens in spring. Our Central Floridian spring starts earlier than up north, but we have the same stock flush in our spring timeframe.

Spring stock flush

For Central Florida, these timeframes are:

March 1st – Best spring selection.

November 1st – Winter selection starts but it is not flush like spring.

One thing to note is that the best selection occurs just as we are heading into our dry period!

Big box stores, which have a higher stock turnover, will have a big spring flush. Beyond that, they are very cyclical. Plants are brought in when in bloom, and once they sell out, they move on to other plants.

Independent garden centers will have the flush too, but will also have a better selection year-round because they always need plants for landscape installations. However, they will cut back in winter because otherwise, they must cover much of their stock, a time-consuming process.

Part 2 - Implementation and Care

SELECTING SPECIMENS

BUYING PLANTS

When buying plants, there are several things to watch for. Inspect the leaves for insects and shake the plant gently to see if any fly away.

Look at the bottom of the pot to see if roots are growing out of it, in which case the plant is already root-bound. Circling roots in a pot usually continue to circle after being placed in the ground. If it is an anchor root (versus tiny feeder roots), it needs to go straight down into the soil to provide stability for the plant. Circling roots can slow a plant's growth or kill it altogether.

Overgrown roots

Circling root

When choosing trees, pick one with a strong central leader (trunk), which helps to withstand hurricanes.

Garden centers may keep plants around on their shelves until they sell. So, for annual flowers, be sure it is the proper growing season, and for your best value, at the beginning of the season and not the end.

Strong central leader

Split central leader

PROPAGATING PLANTS

You can also have fun propagating to create your own plants. However, this takes time, and it is probably just easier and cheaper to buy most plants.

SEEDING

Most landscape plants are not seeded, but seeding is typically done for annual flowers. Just because you find seed packets for a given flower in the seed store does not mean that the flower species is meant for Central Florida.

DIVISION

Grasses and some grass-like flowers grow in clumps and can be divided at the crown. This is done in early spring. Stick a shovel into the middle of the plant, then remove half to another location to grow on there.

CUTTINGS

Many shrubs can be propagated by cuttings. This method does not always have a good success rate. Success depends on a lot of factors beyond the scope of this book. Before you begin, invest in a small jar of rooting hormone, a white powdery substance. This will help roots form. Use vermiculate or a seeding mix for best results.

Cutting

Start by taking a cutting of about 6" from below a leaf node (where the leaves attach to the stem), and where the stem is slightly flexible instead of purely woody. Strip the lower leaves, leaving 2-3 pairs at the tip. Water the end of the cutting, then dip it in the rooting hormone. Make a hole in the mix, then gingerly place the cutting into the hole so the rooting hormone does not rub off. Cover two nodes. Gently push the mix around the cutting to close the hole.

Place your cutting(s) indoors under grow lights, preferably, or in a sunny window. Keep the mix somewhat moist. The plant has successfully rooted when you can gently tug on the cutting and it stays put.

Part 2 - Implementation and Care

LAYERING

This is a little-known technique that can be done directly in the garden. It works best with certain plants, such as hydrangeas.

Find a stem on the main ("mother") plant that is at least 6-8" long. Do not cut it; instead, leave it attached to the mother plant. Strip the leaves from the center of the stem section over several nodes. Take a paperclip or plant stake and place it over the stripped section, then push it gently into the ground so that the soil covers the paperclip by about an inch.

Layering pin

After several weeks, you can try lightly tugging on the end piece, and if it feels secure, it has grown roots at the nodes where the stripped leaves previously were. At that point, you can cut it off from the mother plant, and you have an entirely separate new plant.

HOW TO PLANT

THE SUNSHINE LAW

The sunshine law is also known as the "call before you dig" law. In Florida, the phone number is 811. Before you even think about planting, you need to perform this step. When you call, you will need to identify where in your yard you will be planting - front, back, etc. Usually, within a couple of weeks of calling, the various utility companies, including electric, cable, water, etc., will mark where the utilities are on your property, leaving a colorful artwork of painted lines on your lawn (do not fret, your grass will grow out). Then you or your landscaper can feel comfortable digging without hitting something. I personally know someone who accidentally hit electrical lines (through no fault of his own) and barely lived, so this step is essential.

Installation contractors should do this for you. If installing yourself, you must be the one to call.

DIGGING IN

Dig a hole that is wider than the plant container and deep enough so that the soil line of the plant in the container will be one inch above the ground (it will sink over time). The bottom of the hole should be solid.

Gently remove the plant. Hold the soil in place with one hand and turn the pot sideways or upside down to loosen the plant. You can also roll a round container on the ground, using your palm, to loosen the plant. Be sure to take the entire root system and its soil, trying to keep everything together.

If you see that the plant is root-bound (i.e., roots are very tightly entangled), gently tease (in other words, pull) the roots apart. One day my husband was helping me in the garden. As he pulled a plant out of its pot for putting in the ground, he started talking to it. "Your mother was a nasty weed; your father was poison ivy." Quizzically, I asked what the heck he was doing, and he elaborated, "The plant was root-bound, so I was teasing the roots." Quite the sense of humor, my husband has!

Place the plant in the hole, cover it halfway with soil andwater andfill the remainder with soil. Water thoroughly. Do not tamp down the soil since plant roots need air.

If planting a tree, be sure the trunk flare is visible. If a plant is planted too deeply, it will die a slow death. One time I saw this happening in a hedge row of plants, where just one plant was dying in the middle of the row, but the rest were fine.

Trunk flare properly visible

STAKING

A common myth is that trees need to be staked. In reality, better trunk strength is achieved by letting the trunk sway in the wind. Exceptions are if planting in an exceptionally windy area or if the tree has shallow roots. The tree may need staking so it does not come out of the ground until roots are established. The other exception is single-trunked palms. Palms grow all new roots after planting, so they will need staking until roots grow and stabilize their heavy tops.

Staked palms

Stakes should be removed after one year; otherwise, the ties will strangle the tree as the trunk gains girth. Landscapers may use stakes during the installation process, but I have never heard of one coming back a year later to remove the ties. It is left to the homeowner to do so – if the landscaper even tells them they should.

One time homeowners had recently purchased a house that came with a dying plant. During my consultation with them, I noticed that the stake was still in the ground. Upon further inspection, I saw that the trunk had outgrown the stake tie. The tie was cutting completely through the surrounding bark (where water and nutrients are taken up), strangling the tree to death.

Dying Weeping Bottlebrush

Tie strangulation

Part 2 - Implementation and Care

MOVING PLANTS

There is a 50% chance or less of successfully moving plants. In fact, if you have an existing special shrub that you would like to keep in a plan, landscapers will leave it in place. If you want a group of common plants to remain but in a different location in the new plan, landscapers will probably take the old ones out and put new ones in at the new location. There is too much risk of failure.

There is a way to successfully move a plant, but it takes time, something landscapers are not going to deal with. It involves air pruning by making spade cuts sporadically around the perimeter, letting the plant adjust, making more cuts, letting it adjust, and so forth. If you want to try to move a plant, you can increase your odds as follows.

- **Take the entire root ball**. Most landscapers do not do this, they are used to removing/discarding rather than moving.
- **Rainy season and cool weather are the best times.** The dry season can be the kiss of death; it is just too much stress on the plant. Fall, when the soil is warm but new growth enters dormancy, is best. Spring is the next best time.
- **Water, water, water.**

Part 2 - Implementation and Care

PLANT & LAWN CARE

WATERING PRINCIPLES

IRRIGATION SYSTEMS

Watering efficiently is Florida-friendly landscaping principle #2. It means to irrigate properly to reduce your water bill, prevent extra nutrients from going into the water supply, and reduce fungal diseases.

A rain gauge/sensor is part of an irrigation system and is required by Florida law. It is an important piece, especially with the deluges of rain we can get. It will prevent overwatering by keeping track of how much rain was received or irrigation water used. You can verify that it is working correctly by using a hose to fill it and then running your irrigation system to ensure it does not kick in. The sensors may need to be replaced every few years.

An irrigation system should have regular maintenance to keep it in working order. Maintenance includes cleaning or replacing broken heads, testing your rain gauge, and changing your timer seasonally. You can do maintenance yourself or hire an irrigation contractor to inspect your system quarterly.

Many communities use reclaimed water for irrigation, such as from retention ponds, which helps with sustainability.

HOW OFTEN TO WATER

How often you should water will vary by season. There are differing opinions, however, on how often to water during each season. The University of Florida (UFL), after doing research testing, promotes watering less often, with the view that it will cause roots to seek deeper ground that has more moisture, thus making them stronger. Landscape maintenance contractors I have spoken with, however, say they see better results in homeowner situations with slightly more water.

General guidelines are:
- **Warm season:** 2 (UFL) to 3 (landscapers) times per week.
- **Cool season:** 1 (UFL) to 2 (landscapers) times per week.

I think a good compromise that makes excellent sense because it is based on temperature (which determines stress on a yard) is a schedule one lawn contractor put forth to their customers:

- **60° or below** - No need to water.
- **70's** - Water 1 time per week.
- **80's** - Water 2 times per week.
- **90's** - Water 3 times per week.

As far as when to water each day, 2-7 a.m. is the best time. It mimics morning dew. And it allows foliage to dry before nightfall, helping to reduce fungal diseases.

The days of the week a property is allowed to water may be specifically designated and limited.

HOW MUCH TO WATER

Regardless of how often watering is needed, most experts agree on the amount, which is ½" to ¾ "per watering event.

If you suspect a problem in your yard, do the "tuna can test" to verify. Place tuna or pet food cans or deli containers around your yard, including areas that are doing well and troublesome spots. Let your irrigation run as usual, then measure. If some areas are watered less or too much, you may need to try different irrigation heads or have your irrigation system revised to ensure full coverage.

Tuna can test

Irrigation systems are created with different zones. The number of zones will vary by the size of your yard, but lawns will be on different zones than garden beds. Lawns require more water than plant beds. For example, your lawn may need 45 minutes per zone, whereas your garden beds may only need 10 minutes.

The quality of your soil can influence how much water you need. Compost, for example, helps to maintain moisture. Organic mulch also helps to retain moisture.

WATERING PLANTS

IRRIGATION SYSTEMS

Watering of lawns is done via overhead irrigation from ground-level sprinkler heads. The best way to water plants in garden beds, however, is to water the roots rather than the foliage, which is accomplished via drip irrigation. This helps reduce fungal diseases, reduces weeds, and reduces the amount of water needed. Emitters can clog occasionally. But many homes have overhead irrigation that came with their property, for not just their lawns, but also their garden beds. These are generally taller sprinkler heads.

One problem I see frequently in garden beds is that inappropriate shrubs, such as drought-resistant plants or those susceptible to fungal diseases, are placed too close to an irrigation head. Also, as plants mature and get taller, they may block the irrigation head from getting water to other plants.

If you have recently added a new garden bed, your irrigation piping should be adjusted. Otherwise, you will need to decide – do you set the zone area to lawn needs, in which case your new garden bed will be overwatered, or do you set it for plant needs, in which case your lawn may not get enough water?

PLANT WATERING SYMPTOMS

Depending on your specific yard and location, and the types of plants you are using, you may not even need to water plants in Central Florida at all except during dry spells. In most communities though, irrigation systems are installed with every new home.

Symptoms of underwatering and overwatering in plants are very similar. Yellow leaves that detach easily and have browning leaf tips and margins are generally due to underwatering.

Yellowing green leaves with brown tips that stay attached and are somewhat droopy or soft are usually due to overwatering. Round brown legions, many with a yellow halo, are fungal and result from too much water. If sun plants are placed in shade, they may not be able to use all the moisture and will have overwatering symptoms. Poor drainage can also cause problems.

WATERING NEW PLANTS

If you have an entirely new landscape, a 30-day grow-in period of daily irrigation is allowed. But if you have only added a new garden bed or new individual plants, you will have to hand water initially until the plants are established so that you are only watering the new plants and not impacting the existing areas.

When we bought our current home, the builders came down to the wire with the closing deadline, and the landscape was installed only two days before closing. They were not able to get the 30-day water-in period as a result and set us up with just the normal irrigation schedule. Luckily it was in winter, when plants are dormant anyhow, but two of our plants did die and had to be replaced.

The amount of water for new plants depends on the season they were planted in (warm/cool, rainy/dry). Another factor is the tree trunk diameter or shrub size of the container. Usually, independent garden centers or your installer can give you a schedule. A frequent recommendation for shrubs is watering once a day for the first week, every other day for the second week, etc. For trees and palms, it is generally based on the caliper (or diameter) of the trunk – a small hose running for 20 minutes daily for the first month, every other day for the next month, etc.

It takes longer than you think for a new plant to become established – approximately 6 months for shrubs and a year for trees. Do not rely solely on your irrigation during this time, especially in droughts, as it will not provide enough moisture.

Landscapers will frequently install bubblers on new trees and palms. These specialized irrigation heads provide more water than normal heads. You should keep them on for a year until the tree or palm gets established, and then turn them off.

FERTILIZING PLANTS

Florida-friendly landscaping principle #3 specifies fertilizing appropriately, including using proper amounts and following a proper schedule.

FERTILIZER COMPONENTS

Fertilizers include major nutrients called macronutrients, known as N-P-K.

- **N = Nitrogen** - Nitrogen helps plant foliage and grass.
- **P = Phosphorus** - Phosphorus is for plants that produce flowers (not lawns). Florida soils already have plenty of phosphorus naturally, and too much is a leading cause of our lakes and waterways becoming polluted.
- **K = Potassium** - Potassium is used for root health, including overall plant health and disease prevention, for both plants and grass.

Fertilizers vary in how much of each of those major nutrients they contain. For instance, a fertilizer listed as 10-10-10 is considered a balanced fertilizer and has equal proportions of all the major nutrients.

Fertilizers may also contain micronutrients, such as calcium and magnesium, as well as other ingredients, such as vitamins, probiotics, and mycorrhizal fungi. Fertilizers come in two types – liquid for a quick hit or granular for slow-release.

HOW OFTEN AND WHICH TO USE

Overfertilization will result in brown leaf margins and root dieback. It will also increase pest problems since pests love to feed on tender new growth. And it will require more water due to the rapid growth.

If a plant already has a deficiency from under-fertilization, it may not correct existing foliage but will improve new foliage (except nitrogen).

Below is a schedule and the type of fertilizer to use for various plants.

- **Trees and shrubs.** Most trees and shrubs, once established, do not need to be fertilized at all. Think about all those plants in the woods! For plants that are young, or to improve flowering, fertilize every 3 months during the growing season – March, June, Sept.
- **Roses and citrus.** These plants have their own unique needs and have their own fertilizers that perform best for them. Follow the same schedule as for normal trees and shrubs.
- **Acidic plants.** Some plants need a more acidic soil to thrive, such as Azaleas. There are specialized fertilizers for them, sometimes noted as "for acidic plants," other times as "Azalea food." You can find the list of plants that would benefit from it on the package label. Fertilize according to the regular schedule as normal trees and shrubs. If your plants have problems, it could be that your acid-loving plants were placed too close to your concrete house foundation.
- **Palms.** Palms need special fertilizer; see the Palm Care section.

Many Florida counties, including Lake and Marion, have adopted an environmentally friendly ordinance to protect our water bodies and aquifers. For instance, Lake County in Central Florida, due to all its natural lakes as the name implies, restricts fertilizing from June 1st to September 30th. Many counties also limit the amount of nitrogen that can be used and state that it must be slow-release.

One important point—Landscape maintenance contractors use lawn fertilizers and general garden bed fertilizers. However, I have yet to find one that will apply specialized fertilizers, such as for acidic plants. And no, you cannot add those yourself on top of what they already applied because overfertilizing can kill a plant. How to get around it? Hire someone for your lawn, but fertilize trees and shrubs yourself.

FERTILIZING NEW PLANTS

Do not apply fertilizer on newly planted areas right away; rather, wait 1-2 months. It will burn tiny feeder roots in the top 2" of soil trying to get established. And many growers have put fertilizer in the pot already.

HOW TO APPLY

Apply fertilizer starting 6-12" out from the trunk/stem, and fertilize all the way out to the edge of the canopy, called the canopy drip line. Someone once asked me to look at their Hibiscus that was brown and dying. Initially, I did not see anything unusual until I looked at the ground and found a very large pile of fertilizer around the trunk. Although I attempted to remove it, the damage had been done, and it did not make it.

Canopy drip line

Fertilizer needs moisture and can pull it from plants or grass. Hence, water it in after applying. ¼ inch is fine, up to ¾" maximum, because more than that will merely leech the fertilizer right through the soil.

What is wrong with the scenario in this photo on right? The fertilizer area (the canopy of the tree) overlaps the lawn because the block wall was made way too small. It would be okay if it were a palm tree, which can use the same fertilizer as a lawn, but in this case, this flowering tree will not be able to get enough of the proper fertilizer.

Canopy drip line extends into the lawn

Palm canopy okay

Part 2 - Implementation and Care

PRUNING PLANTS

Pruning is a maintenance chore, and you can cut down on much of it by choosing plants properly sized for the space.

WHEN TO PRUNE

Pruning timeframes are different in Central Florida than in other parts of the country.

- **Diseased** – Anytime.
- **Non-flowering deciduous trees** – In winter while dormant.
- **Roses** – Mid-February (Valentine's Day).
- **All others** - March 15th (our last frost date) through October 15th (when plants enter dormancy).
- **Flowering trees/shrubs** – After flowering has completed.
- **Fast-growing plants** - May need pruning every 3 months.

Landscape maintenance companies may try to sell you a package for regular pruning every 3 months. Do not fall for it. If your shrubs are pruned in the fall, they will want to put out new growth, which prevents them from going into dormancy. And that new growth could get zapped in a freeze and thus damage a plant.

PRUNING STYLES

Which do you prefer – a natural-looking landscape or one that has been pruned extensively? Properly sized plants will not need to be pruned and, therefore, will automatically have a natural look. If you wish to trim back and keep that natural aspect, use hand pruners to do so.

Lovely natural look

Practical Landscaping

Landscape maintenance contractors will use a hedge trimmer (they will not take the time to hand prune) to create uniform mounds. I have seen many normally beautiful shrubs pruned in this manner that end up looking utterly destroyed. Eventually they will probably grow out, but only if left to do so, since many maintenance companies prune on a egularly set schedule.

Pruned mounds

Sometimes I see what I call lollipop shapes, which is when the lower foliage is trimmed from the bottom of a shrub, or a tree is pruned abnormally. It looks very comical and unnatural. I suspect the reason is that fungal diseases come from the soil and can splash upwards. But with mulch or rock, this should prevent it for the most part. Large Ligustrum shrubs are pruned up because they are very prone to fungal diseases, but the way it is done is natural-looking.

Lollipop tree

Trimmed up Ligustrum

Sometimes pruning becomes a work of art. Beautiful shapes are created in greenery. If you have a special tree that needs to be pruned, you would be better off calling a certified arborist.

Fancy pruning artwork

Part 2 - Implementation and Care

PROPER PRUNING

The general rule of thumb for pruning is this – never prune more than a third of the plant. More than that will create stress and the plant may fail. There are some plants that can take a heavier pruning, even to the ground, but unless you know for sure which ones, follow that general rule.

Never prune more than 1/3

Improperly pruned hedge

Hedges are okay to trim with a hedge trimmer. The proper way to trim is in a triangle shape so the bottom is wider than the top. It allows the bottom to get sun, which prevents plants from getting leggy. I have yet to see it done anywhere properly.

While pruning, if a plant oozes a milky substance, it could irritate the skin. It is best to wear gloves in those situations.

Pruning techniques vary by plant type, which is beyond the scope of this book. However, here are some quick tips. Remove any dead growth from the interior of a plant, to increase air circulation, important in our humidity. Crossing branches within a shrub or on a tree should be removed. Making a cut b ack to the trunk (but not too close for injury) will permanently remove the branch. Cutting back to above a node (where the branch intersects the trunk or another br anch) will encourage new growth. Ornamental grasses are usually cut back in late winter, before new spring growth starts, to 6-12" tall. (For proper timing, watch when your community areas are done.)

Pruned grasses

CREPE MURDER

Landscape maintenance contractors will sometimes prune Crepe Myrtle branches back to nubs in the fall. This practice is commonly called "crepe murder." I asked a contactor one time why this was done and was told it was to prevent disease. Years ago, Crepe Myrtle trees did have more diseases, but nowadays, many are resistant. Regardless, this practice sh ould not be done because it weakens limbs over time, inviting disease.

Crepe murder | Crepe Myrtle growing out | Crepe Myrtle in decline

I once was at a big box store and saw that an entire row of trees in the parking lot islands had been cut in that manner. The problem was that I was there in the summer when Crepe Myrtles should be in peak bloom season, and yet, the entire row of trees did not even have leaves. Such a sad sight to see them all dead.

Crepe Myrtles come in different sizes, from smaller shrubs to large trees, so choose the correct one for your space. If you inherited one when you bought your property, and you feel the need to trim, follow the standard rule of thumb to reduce the tree height or hire a licensed arborist who will know how to do it properly.

Part 2 - Implementation and Care

CONTROLLING TREE HEIGHT

Here is a scenario. A nail is placed in a tree at 1' high off the ground when it is 1 year old. The tree grows 1' each year. How high is the nail at 5 years old? The answer is – the nail is still at 1' because growth comes from the top.

Knowing that, you can control the growth of a tree with a central leader (trunk), by cutting off the top when it has reached your desired height. Also, if you cut lower limbs, they are gone forever.

Holly central leader cut

REVERSIONS

Sometimes plant varieties will put out branches that will revert to the main plant color/type from which they were originally developed. For example, variegated plants were created from plain green plants originally, and every so often, they will put out a green branch. Just prune it out.

Reversion back to green on variegated Pittosporum

Sometimes new growth on a plant will be different than the base color. For example, if you prune yellow foliage plants, you are apt to prune off the yellow and be left with green since yellow is new growth only. But do not worry, it will return.

Green undergrowth on yellow Sunshine Ligustrum

Sometimes a new type of offshoot is created spontaneously by nature, and that becomes a "sport." Pruning and propagating it can create a rare plant that may become a collector item and, in turn, make you rich. Despite my many years of growing plants and looking for them, I have yet to see this happen in my yards. But I have bought plants resulting from sports.

TREE HANGING PLANTS

In Florida we have beautiful moss hanging from our large oak trees, which signals to us and visitors that we are in the true south. You are apt to see:

- **Spanish moss** – A native air plant that uses the tree for support, but takes its nutrients from the air and not from the tree in any manner.

Spanish moss

- **Ball moss** – Another type of native air plant.

Ball moss

- **Moss/lichen** – Found on tree trunks and more prevalent on some species than others. It can be a result of shade, particularly on the north side of a plant, and water.

Lichen on Maple tree trunk

Homeowners get concerned seeing these on their landscape trees. None of these are harmful. They are not hurting the tree, merely using the tree to hang on to, so there is no need to remove them.

WEEDING BEDS

TIMING OF WEEDS

Weeds are classified as either annuals (meaning they complete their life cycles in one season) or perennials (year-round). For annuals, they are classified as summer/warm season (completing their life cycle from spring to fall) or winter/cool season (completing their life cycle from fall to spring). Identifying which weed you have makes a difference as to when you need to spray for control.

TYPES OF WEEDS

There are three types of weeds:
- **Broadleaf**
- **Grass** (round hollow stems)
- **Sedge** (has triangular-shaped solid stems)

Sedge grass

It is important to know what kind of weed you have so you can apply the proper chemical. If you are spraying for weeds in your garden bed, a grass weed chemical is fine (unless you have ornamental grasses). But if you use a broadleaf chemical, be careful not to spray too close to a plant because the chemical does not know the difference biologically between a broadleaf weed and a desirable plant.

WEEDING CONTRACTORS

And where do most of the weeds form? Around the base of a plant, exactly where it is not safe to spray. Landscape maintenance contractors will only spray, they will not hand-pull at all or else only do so if the weed is higher than the plant. It is almost impossible to find someone who will hand-pull weeds for you.

Many will not perform weeding in lawns, because it is extremely hard to maintain, and there are many complaints. They usually perform service once a month, and we all know weeds in Florida grow much faster than that.

Practical Landscaping

COMMON WEEDS

Weeds can be indicator plants, meaning certain weeds grow in certain conditions, which could help you identify problems in your yard. These are some examples of common weeds in Central Florida, their timing, and what they may mean for your yard.

- **Artillery Fern** - A perennial succulent that likes moist disturbed areas and will grow anywhere. It spreads rapidly via ejected seed.
- **Chamberbitter** - A summer annual.
- **Chickweed** - A cool season annual that likes compacted soils or moist conditions. It can also thrive where lawns are mowed too low.
- **Crab Grass** - A warm season annual that likes acidic soil low in fertility.
- **Creeping Woodsorrel** - A perennial with red leaves and yellow flowers.
- **Florida Pusley** - A warm season annual that likes dry soils and can also be where nematodes thrive. It has a deep tap root that makes it difficult to kill the first time.
- **Hawksbeard** - An annual with a tall small yellow flower.
- **Sedges** (yellow and purple varieties) - Perennials that love moisture, so check if you are overwatering.
- **Spurges** - Include 3 types – spotted (prostate), garden, and hyssop (upright). They have red stems that are milky inside when broken. They are warm-season annuals that like dry or compacted soils.

Artillery fern

Chamberbitter

Florida Pusley

Spurge, garden

Spurge, hyssop

- **Yellow Woodsorrel** (many times mistakenly called clover) - A perennial weed, with pods that explode seeds 15' away. It likes alkaline soil.

CONTROLLING WEEDS

Weeds are a foregone conclusion to any yard. They end up in lawns and garden beds, mostly coming from the wind, your lawnmower person from the prior client's yard, birds pooping seeds, and more. Keeping on top of them will prevent them from spreading and making the problem worse. Here are some suggestions.

- **Apply a pre-emergent product.** Pre-emergent products prevent weed seeds from germinating. Organic products exist also. Products should be applied in mid-February to prevent summer weeds, and in late October for winter weeds.
- Do not bother using weed 'n feed products in Central Florida or you are wasting your money. It is not proper timing for us – February is for pre-emergent weeds, April for fertilizing. So, if you use such a product in April, you missed the window for weed prevention, and if you use it in February, the fertilizer will be long gone by the time plants can actually use it.
- **Apply a post-emergent product.** If the weed is already there, you will need to use a post-emergent weed killer product/spray or hand-pull.
- **Add mulch.** If you have rock mulch, you cannot do this, but if you are using organic mulch, continually add it every few years to prevent weed seeds from germinating.
- **Stop digging.** Every time you dig to replace a plant or add more plants, weed seeds are brought to the surface. After you have planted, watch for new seedlings. Even hand-pulling weeds will disturb any seeds in the soil. Place your fingers around the weed before gently pulling, thereby keeping the soil as undisturbed as possible. Weeding when soil is moist may also help.
- **Remove flowering weeds.** If you do not have time to weed, remove or spray the ones that are in flower immediately, and come back and do the rest another time. If they are in flower, they are going to set seed and magnify your weed problems.

ROCK MULCH WEEDS

Landscape fabric is installed over the soil first, to keep the pebbles from sinking into the soil. Many people assume this helps to prevent weeds (sometimes the fabric is referred to as weedblock), but that is a myth. Yes, it may prevent any existing weed seeds in your soil from germinating, but most weeds come from above. And once the weed roots get a hold in the fabric, they can be difficult to pull out. It is actually easier to pull weeds directly from the soil through organic mulches.

Weeds in rock mulch

Part 2 - Implementation and Care

HANDLING FREEZES

PREPARING FOR A FREEZE

We get light frosts here in Central Florida, and occasionally a hard freeze. A hard freeze is defined as being below 28° for more than 4 hours. Plants acclimate to gradually cooling temperatures as winter approaches, so sudden freezes are more damaging.

Covering your tropical plants will help them get through a freeze by trapping warmth from the earth. However, it must be done the proper way. Cover plants to the ground with sheets or blankets, which traps in earth heat. Do not use plastic since it actually takes heat away from a plant.

Proper covering

Improper covering

"Frost blankets" can be found at nurseries or big box stores. Be sure to purchase in fall, since by waiting until a couple of days before a freeze, they will all be gone. Frost blankets are lightweight cloth, usually made of white polystyrene material. They are sold in different sizes and as a single sheet to cover multiple plants or as individual plant bags. Thickness varies, with most at .9oz or 1.5oz (the latter being more expensive). They can add an additional 5-10 degrees of protection, depending upon thickness. Remove them the next morning.

Watering the day before a potential freeze is coming will keep the soil warmer. Turn off your irrigation that night, however, so leaves do not freeze.

Practical Landscaping

HANDLING DAMAGE

Damage will show as browned or burned-out foliage or as leaf drop. Interestingly, you may think you got through a freeze fine, and then your plants will start having problems 2-4 weeks later. Freeze damage does not always show immediately. The plant will struggle or will not put out new growth or flower buds.

Freeze damaged Roebellini palm   *Freeze damaged Hibiscus*

The hardest part of handling the damage is this – do not prune until normal spring pruning dates. Pruning encourages new growth and should we get another freeze, it will be zapped and cause even further damage. It is hard, but try to wait.

Part 2 - Implementation and Care

PLANT PESTS

Florida-friendly landscaping principle #6 is to manage yard pests responsibly. Primarily, this means following the IPM approach to spraying.

IPM

IPM stands for Integrated Pest Management, and is used for best results while also balancing a healthy environment. It means starting with the least toxic approach first, then moving up to more caustic chemicals as needed. Regularly perform monitoring before spraying. Be sure to check the undersides of leaves also, where pests like to hide or lay their eggs.

Make sure you identify the pest properly before spraying, so you do not accidentally kill a good insect instead of a bad one. Take a photo and use one of the many phone apps these days that can identify a pest. Be especially careful when plants are in bloom because flowers attract bees and other pollinators.

Do not hire a company that will automatically spray your plants on a regularly set schedule. That is not Florida-friendly, and they can needlessly kill the good bugs along with the bad bugs. Many sprays are contact sprays, which means they do not work preventatively, but only if the bugs are already there.

Of course, there are exceptions when it comes to your home rather than your shrubs. Keep up with spraying your home for termites and cockroaches.

The good news is that a healthy plant will resist many problems. In fact, most plants can withstand a 10-20% loss of leaves without major problems. Many bad insects are food for beneficial insects and are kept under control by maintaining a healthy habitat.

COMMON PESTS

The most common insects you are apt to encounter are below.

- **Aphids** – These are green or dark-colored tiny insects that usually hang out on the underside of leaves. They produce a sticky black substance called sooty mold that ants love. So, if you see ants, check for aphids. And remove the sooty mold by spraying with water, to not block photosynthesis

Aphids

- **Whiteflies** – Just as the name says, these are tiny white flies that, when you shake the plant, will fly away.

- **Mites** – Too tiny to see without a powerful hand lens, but you may see their webbing or the result of their feeding via a stippled mottling on the leaves.

- **Thrips** – You will have to look very closely to see these insects that look like a tiny sliver. They are usually found on succulent new growth and cause curling. Their feeding may result in speckled silvery patches on the leaves.

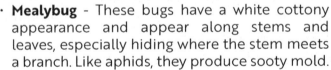
Small deformed new growth on Chase tree

- **Mealybug** - These bugs have a white cottony appearance and appear along stems and leaves, especially hiding where the stem meets a branch. Like aphids, they produce sooty mold.

- **Cottony Cushion Scale** – Scales are found on stems and mid-ribs of leaves, and are usually indicative of an air circulation issue. They also produce sooty mold.

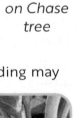
Mealybug on Hibiscus bud

- **Nematodes** – These are microscopic worms that live in the soil. There are good nematodes and bad, and Florida is prone to the bad root knot nematode. They are hard to diagnose and hard to get rid of (other than solarizing, which will kill grass and plants as well). Stunted plant growth can be a sign of nematodes. Compost helps keep them at bay.

- **Beetles** – They chew leaves starting from the outside margins. Use Neem or horticultural oil.

- **Caterpillars** – They leave holes in the center of a leaf. Use Bt.
- **Slugs/snails** – They like to come out at night, but you can usually see their silvery trails in the daylight. They also leave ragged holes in leaves, starting from the base of the plant and working their way up. There are many organic granular products or fun methods (salt melts them) to try on your own.

CONTROLLING PESTS

Keep spraying for several weeks, throughout the insect's life cycle. Make sure to spray underneath the leaves also, where pests like to hide. Some controls include:

- **Insecticidal soap** - Relatively safe, usually organic, and kills smaller soft-bodied insects on contact. It is useful for aphids, whiteflies, spider mites, and thrips.
- **Neem oil** - Organic oil that kills by smothering pests such as mealy bugs. Do not apply it during the heat of the day or it can fry/burn leaves.
- **Horticultural oil** – Synthetic version of above.
- **Miticide** – Specifically kills mites. (Most insecticides work on the nervous system and will not kill mites.)
- **Bt** – An organic beneficial microbial product that can be used for caterpillars and worms.
- **Spinosad** - Another organic product made from naturally occurring microbial bacteria in soils, is effective at killing many different pests.

ANIMALS AND MORE

- **Rabbits** – Rabbits love some of our Floridian plants and can easily decimate them. Fencing or chicken wire works. Be sure to dig it down one foot so they cannot get underneath. Unfortunately, fencing is not allowed in some communities, but if you use a black wire one, it usually cannot be seen from afar.

You can also try growing garlic or use garlic clips on your garden perimeter. There are repellents, including natural red pepper, but they need to be reapplied constantly after rains.

Practical Landscaping

Plastic snakes, surprisingly, seem to be very effective, perhaps because rabbits have bad eyesight? And if you find a real snake skin, leave it!

Our fake snakes keep rabbits at bay, but inadvertently scare our maintenance contractors!

- **Snakes** –Most snakes found in garden beds are black racers or garden snakes, which are good guys, however scary. We found one in our master bathroom one time (my husband removed it safely outdoors). We do have several venomous snakes in Central Florida as well. If you do not want snakes, reconsider using bird feeders (seed attracts rats, rats attract snakes). Also limit large grasses, which provide nesting habitat for rats, and then snakes.

- **Others** - We have scorpions, which sting but are non-venomous in Central Florida (which we found in our master bedroom closet one time). And Black Widow and Brown Recluse venomous spiders (the former of which we found against our house once), but those are another story. And don't you just love the little anoles, especially the green native ones, who eat cockroaches in the landscape.

Good black snake (found in our master bath and safely moved outside!)

Teal garden snake

My husband's shadow over a coachwhip snake (largest in U.S.)

Green anole

Part 2 - Implementation and Care

PLANT DISEASES

DISEASE CATEGORIES

Diseases fall into three categories: fungal, bacterial, and viral. Accurate diagnosing of bacterial or viral can be difficult without lab tests or a high-powered microscope.

Fungal symptoms can usually be seen, which appear as dark spots on foliage. It is caused by too much water, either from overhead irrigation or our rainy season. There are many fungal chemical sprays on the market to help.

Fungal disease

Viral issues are often confused with fertilizer deficiencies. The difference is that viral problems have a non-uniform pattern on the leaves, whereas nutrient problems are symmetrical.

Nutrient Deficiency

OTHER PROBLEMS

Galls are abnormal growths on a tree or shrub whereby the plant walls off the disease, pest, or injury. Normally they are not serious.

If multiple things are going on at once, such as browning leaves, pests, and nutrient deficiencies, it could be due to something at the root level, including being planted too deep.

Very large gall

If you do not know what something is, bring a sample or photo to your County Extension, and they may be able to help.

PALM CARE

Palms are specialized plants, and are in the grass family. Everyone wants a tropical palm when they first arrive in Florida! But over time, people may prefer less work, enjoying their neighbors' palms instead.

WATERING OF PALMS

Follow the same watering guidelines as in the plant watering section above. Cracks in palms, especially Queens, could be a sign of overwatering, which is not always a problem due to the interior vascular system of a palm.

FERTILIZING PALMS

Because of our sandy soil in Central Florida, nutrient deficiencies can occur. Potassium deficiency is common, which is the yellowing of lower leaves with translucent spots and brown tips. Magnesium deficiency is also quite common, with bright yellow leaves with green centers, and it typically occurs with a potassium deficiency. A boron deficiency will cause a bent top and wrinkled leaves.

Palm yellowing

Fertilize palms every 3 months during the growing season – March, June, Sept. The exception is our Florida native Sabal palm, which does not need any fertilizer.

Use a fertilizer meant for palms, but be sure it provides micronutrients or has a designation of "+4", which will include magnesium and other needed palm nutrients. Palm fertilizers can also be used on lawns because palms are in the grass family.

PRUNING PALMS

Palms are either self-pruning, meaning they drop their old leaves, or they need to be pruned. Do not trim leaves on palm trees unless they are totally brown. Palms have an ability known as "translocation," which means they pull nutrients from lower leaves. If you trim those lower leaves because they are yellowing, the palm will just pull from the next layer up, leaving that one yellow.

Over-pruned palm

Seeds of the Pindo palm

Palms can have seed stalks on females. They can be removed, but wait until they are mature, or they will just form another one. Birds like the seeds, so you can keep them unless they are a nuisance or near a walkway.

Some palms have ferns that grow in the boots, and some claim that rats inhabit them. They can be pulled out, but gently so as to not damage the boots.

Some contractors will do what is called a "diamond cut," where the boots are cut back close to the trunk, which is a pretty look. On a Sylvester palm, it brings out the beautiful orange trunk color. However, if not done properly, it can introduce pests and diseases.

Palm boots with ferns

PALM DISEASES

Sometimes roots appear above ground at the base of the palm or several feet up the trunk, even causing the bark to flare out. This is referred to as "root initiation." This is normal and not a disease.

Young leaves may have white marks called "scurf" that resemble scale insects, but this is also normal and they will fall off as leaves age. A palm leaf skeletonizer will leave scraped brown marks on lower leaves, but this is merely cosmetic.

Root initiation

Palms have their own unique disease challenges. Palms do not increase girth via new bark like trees, and therefore, injuries do not heal over. If the core of the canopy still has new growth from its center, there is hope for saving your palm. Some of the more common diseases in my community include:

- **Leaf spots** – Yellow halo- like spots is a fungal disease and occur from overwatering on palm leaves. Back off on watering and apply a fungicide.
- **Fusarium Wilt** – Fast spreading, will kill within a few weeks to a couple of months. A red stripe appears on the petiole (stem base), and one side of the leaf turns brown. It starts on lower leaves and moves upward. It especially occurs on Queen Palms. Do not replant the same type of palm in its place.
- **Lethal Bronzing** – A fatal bacterial disease caused by an insect. The palm looks drought-stricken with bronze leaf tips. Palms will die within 3 months.
- **Thielaviopsis Trunk Rot** – Fungus is introduced into a wound, which eventually causes rot from the inside outward. The trunk eventually collapses on itself, killing the palm. Oozing from the wound can be seen. This can occur with improper pruning of leaves too close to the trunk, especially during our rainy season.

Trunk rot occurred during our rainy season, when fungal diseases are more apt to occur. In this case, the fronds were probably previously cut too close to the trunk. The poor homeowner graciously let me take these photos for educational purposes.

Trunk rot ooze

Part 2 - Implementation and Care

LAWN CARE

Grass is indeed a plant, and it is covered in a separate section here, given its unique considerations.

TYPES OF GRASS

In my community, there are three types of grass used for lawns and another used for fields.

- **St. Augustine 'Floratam'** – An older variety, this is used in the northern sections of my community. It has thick blades, and if you are new to Florida, you may think it resembles crab grass. It does best in full sun and is not always very drought-tolerant.

Floratam

- **St. Augustine 'ProVista'** – A new variety that is resistant to glysophate weed killer products, thereby making weed spray easier. You can test if you have this variety by spraying a small indiscreet area of your yard, and if it turns yellow, but does not die outright, it is probably this variety. It is more bluish-green than 'Floratam.' 'ProVista' is slower growing than 'Floratam' and needs less frequent mowing, is semi-shade tolerant, and is more tightly packed for fewer weeds. However, it is also susceptible to Take-all Root Rot disease and may have more chinchbug issues. In my community, it is used in southern sections, but some areas have resorted back to 'Floratam.'

ProVista

- **Zoysia 'Empire'** – A finer, softer grass blade compared to St. Augustine. It is semi-shade and drought-tolerant. It will turn brown when dormant. It is used in central sections of my community.
- **Bahia** – Bahia is typically used in fields because it can be seeded (versus sod for other grass types) and needs little care. It is brown when dormant, but greens up when it rains.

Zoysia

LAWN WATERING

Follow the guuidelines in the Watering Principles section earlier. If you see footprints left on your lawn after walking on it, if the grass has turned bluish-grey, or if leaf blades are folded, your lawn is underwatered. If your lawn recovers in our rainy season, it is a sure sign of underwatering. Mushrooms are a sign of too much water (which can be hard to avoid during our rainy season).

Mushrooms in lawn during our rainy season

Too much water

Brown spots may be a watering issue, either under or over, or indicative of another problem. If you suspect a problem, do the "tuna can test."

Be sure to check your irrigation system quarterly. Sometimes irrigation "donuts," green cement circles, are placed around heads in lawns to protect them from lawnmowers. It also makes maintenance easier because you can find the low heads that are covered by grass. They do occasionally break, and you will regularly need to remove any grass that has grown inside the circle.

Irrigation donut

LAWN FERTILIZING

What is wrong with the brown lawn in the photo? Nothing! The key is when this photo was taken, which was in February. Browning out is natural because lawns go dormant over winter. The other property in the photo was using fertilizer when they should not have been. By keeping their lawn green during the time it needs to rest, it is under stress and more vulnerable to issues.

What is wrong with this brown lawn?

Wait until April to apply lawn fertilizers. Zoysia grass is slower to green up again in the spring than St. Augustine. Fertilize 2-4 times per year, through October. As with plants, many counties have adopted an environmentally friendly ordinance to protect our water supplies and limit when fertilizer can be applied and in what form. In Lake County, for example, lawns cannot be fertilized (with nitrogen and phosphorus) between June 1st and September 30th, and for the remainder of the year, nitrogen must be 50% or more in slow-release form.

Slow release is better on lawns, although lawn maintenance contractors may add liquid in early spring for a jump start. Lawns do not need a fertilizer with a middle number other than 0, and generally, a 15-0-15 fertilizer is recommended. Iron can also be used to assist with greening up (and will not add extra growth like nitrogen). Fertilizers specifically meant for lawns will be labeled as such.

After applying, water in with ¾" water, any more will just go through the soil immediately.

MOWING

Lawns need to be mowed once growth starts in the spring and usually every other week initially, and then as we progress into summer, mowed every week until growth slows again in the fall.

When we first arrived in Florida, my husband and I would laugh at the landscape maintenance contractors mowing in December. What were they mowing, and why? Because people pay them year-round, we guessed they needed to show they were doing something for the money. The only thing growing in your lawn in the winter is winter weeds. However, it may be difficult to find someone who only mows in the warm season.

It is important to follow mowing height recommendations because not doing so is one of the major ways to introduce disease.

- **St. Augustine** (any variety): 3½-4 inches high.
- **Zoysia**: 2-2½ inches high.

TRIMMING

Trimming is done to keep grass out of garden beds or away from hardscapes. It is done using a lawn trimmer or weed whacker tool.

Be sure they are not used too closely to trunks of trees or shrubs, resulting in damage. If the damage is significant enough, or if it strips bark entirely around the trunk, the plant will not live (water and nutrients are taken up via tissues located just underneath the bark, and it results in their supply being cut off).

Tree bark damage

LAWN WEEDS

Since our dogs use our yard to play and because we believe in a healthy environment, we initially tried to manage our lawn ourselves organically, just like we did back in New England. Unfortunately, it quickly became too much for us to handle, especially with weeds and other problems coming in from the empty field next door to us, and everything going crazy in the heat and rains of summer. As a result, our lawn is something we outsource nowadays. Many lawn contractors will spot-treat for weeds, rather than doing an entire lawn, which is more environmentally sound and better for all the good insects that may otherwise be harmed.

Besides the weeds listed in the Weeding section, one weed that becomes a problem in lawns is invasive common Bermuda grass. It is different from the nice improved Bermuda grass variety used for lawns and golf courses. It has fine blades on opposite

Bermuda grass

sides of the blade stem and spreads obnoxiously by both above-ground stolons or underground rhizomes. (Many people mistakenly think it is torpedo grass, which has alternating leaves pointing upwards like a torpedo, and spreads mostly by pointy underground rhizomes.)

Bermuda grass is a perennial and hard to eliminate because it, too, is a grass, just like your lawn, and chemicals cannot biologically differentiate between the two. Hand-pulling can make the situation worse because new sprouts will form where cut. To get rid of it, you will need to remove that portion of your lawn, including all roots, then spray several times to ensure it is fully removed and eventually re-sodded.

OTHER LAWN CARE

Adding compost to your lawn helps regulate moisture, and may result in the need for less watering. In our sandy soil though, it will need to be reapplied every few years.

Lawns may need to be dethatched once per year. Thatch is dead grass that can smother the soil if it gets too thick. It can be exacerbated by too much nitrogen or water, or by using an incorrect mowing height. Zoysia grass is more prone to it, but if dethatching St. Augustine, be careful not to pull up roots. Adding compost can add microbes that will decompose the thatch.

If your lawn is compacted, especially after heavy equipment or materials have been placed on it, have it aerated. Aeration is when small plugs are removed to provide for air. This can be done every 2-3 years in spring. If your soil is extremely compacted, it may become hydrophobic, which means water will run off instead of being absorbed.

If you need to replace sections of your lawn, the best method is by using sod, versus seed, and the best time is spring. Many seeds sold are not the correct variety for an existing yard. And re-sodding will give you an immediate impact, which could be important if you live in a community with deed restrictions on how your lawn should look. It will take approximately 6 weeks to knit together and become established. Be sure to give adequate moisture during that time. And do not fertilize for one month, until established.

LAWN PESTS

There are several predominate pests found in Central Florida and they usually have a preference as far as grass type.

CHINCH BUGS

- *Timeframe:* Mar - Sept.
- *Affected:* St. Augustine.
- *Symptoms:* Yellowish grass, then irregular brown spots. May start close to hot surfaces like roads. Water-stressed, over-fertilized, or compacted areas are most affected.
- *Test:* ¼" bugs are hard to see. They like to hide in thatch. Spray the edge of spots with liquid soap to see if they surface.
- *Control:* Rotate pesticide types to effectively control.

HUNTING BILLBUG

- *Timeframe:* Fall – Winter.
- *Affected:* Zoysia.
- *Symptoms:* Irregularly shaped yellow patches caused by a larvae grub or adult weevil.
- *Test:* Grass easily pulls free (the result of grubs feeding). Check for sawdust-like debris inside the grass. Check for white larvae grubs with brown heads in the top 2" of soil. Adult is a 1/3" weevil that eats the blades.
- *Control:* Pesticide.

MOLE CRICKET

- *Timeframe:* May - Aug.
- *Affected:* St. Augustine, Bahia.
- *Symptoms:* Dead patches, with tunnels that look like tire treads.
- *Test:* 1¼" pre-historic looking bugs like light, so take a flashlight out at night to see if they fly.
- *Control:* Pesticide.

NEMATODES

- *Timeframe:* All year.
- *Affected:* St. Augustine, Zoysia.
- *Symptoms:* Scattered yellowing and thinning out; feed on roots, making grass more susceptible to stress and drought.

- **Test:** Roots are short (less water and nutrient uptake) and stubby, and may have galls. Note: The worms themselves are too small to be seen without a microscope.
- **Control:** Reduce compaction. Use compost with beneficial microorganisms to strengthen roots and keep them at bay. Spray with Idemnify (fluopyram) product.

TROPICAL SOD WEBWORM

- *Timeframe:* May - Sept.
- *Affected:* St. Augustine, Zoysia.
- *Symptoms:* Leaf tips stripped of green; webbing on top of grass.
- *Test:* Shuffle through your grass to see 1" white/grey moths flying. May need to use a flashlight at night to see because they hide in thatch during the day. Check for green pellets (poo) on leaf blades.
- *Control:* Pesticide.

TUTTLE MEALYBUG

- *Timeframe:* Summer.
- *Affected:* Zoysia.
- *Symptoms:* Brown dead patches or general dieback.
- *Test:* Look for white wax on blades. Unroll the bottom of a leaf blade and look for a tiny 1/10" pink mealybug. Insect dies over winter.
- *Control:* Difficult to control; try systemic pesticide, rotate types. Remove thatch.

Part 2 - Implementation and Care

LAWN DISEASES

There are several major diseases in Central Florida, usually by grass type. Most can be fixed, but Take-All Root Rot cannot.

DOLLAR SPOT

- *Timeframe:* Nov - May.
- *Affected:* Mostly Bahia, Bermuda, but also St. Augustine, Zoysia.
- *Symptoms:* Small spots less than 3", look like dog pee spots.
- *Test:* Check for lesions on leaf blades of dead spots with brown edges.
- *Control:* Fungicide.

LARGE/BROWN PATCH

- *Timeframe:* Nov - May, temps under 80°.
- *Affected:* St. Augustine, Zoysia.
- *Symptoms:* Distinct circular patches larger than 3" with a copper halo. Caused by too much moisture.
- *Test:* Leaf easily pulls free of the sheath. Sheaths are brown.
- *Control:* Fungicide. Lessen moisture and nitrogen fertilizer.

LEAF SPOT

- *Timeframe:* May - Sept.
- *Affected:* St. Augustine (Grey Leaf Spot), Zoysia (Leaf Spot).
- *Symptoms:* Browning.
- *Test:* Check for small oblong spots on leaf blades. Overwatering?
- *Control:* Fungicide.

TAKE ALL ROOT ROT

- *Timeframe:* Summer.
- *Affected:* St. Augustine, Zoysia.
- *Symptoms:* Irregular patches 8-24" or more of yellow and dead grass. Occurs in stressed areas (too much fertilizer, incorrect watering) or areas that have been mowed too short.
- *Test:* Check for roots that are short, black, and rotted.
- *Control:* Spray with fungicide. May need to totally replace the affected area, but even then, it may not come back, in which case, use a different ground cover instead of grass.

GENERALIZED BROWNING ISSUES

- **Overwatering** (purple blades) – Test via "tuna can" test (place cans around the lawn, run irrigation, measure water). Adjust irrigation or heads, and check for working sensors.
- **Underwatering** (bluish grass which leaves footprints) – Test via "tuna can" test. Adjust irrigation or heads, and check for working sensors.
- **Over-fertilizing** – Reduce the amount of Nitrogen in particular, water immediately after application.
- **Compaction** – Aerate.
- **Scalping** – Use correct mowing height.
- **Thatch** – Dethatch Zoysia every 2-3 years.
- **Dormancy** - Normal for Zoysia in winter and may also occur in stress.

PART 3 - PLANTS

Practical Landscaping

HOW THIS PART WORKS

In this part of the book, you will find the most frequently used landscape plants in our Central Florida area. And – a key point – they can easily be found. Many other resources include plants that may meet zone requirements, but they are not available locally. Many resources also do not have information on specific varieties used in landscapes. For example, there is a big difference between plants that are noted as growing 3-15'! This is some variance in size due to site location and conditions, but not that much.

Included in this book are descriptions of individual plants and when necessary, their varieties. They are categorized by type and then alphabetically. There is a full index at the end. All plants are considered Florida-friendly and are hardy for our zone 9 (or otherwise noted).

The most commonly used plants are noted in blue font. I have also included plants that are not necessarily found in landscapes, either because they get too tall, spread, or are too expensive. But they are still found in community areas or may be used, if you have the space, in specialty gardens. It is always interesting to know what else grows in our locality.

Designations are made as to whether a plant is native or if it attracts wildlife (bees, birds, and butterflies). Wildlife may be attracted due to flowers or berries, or because a plant offers nesting and protection. Many resources indicate that a plant is an attractant, but in reality, it does not appear to be so. Included here are the most common and easily found plants used in landscaping that will indeed attract based on the actual experience of myself and friends.

If you wish to delve further into uncommon plants such as natives or rose varieties not typically used in landscapes, visit a nursery that specializes in those, or participate in local clubs for additional information.

Key is as follows:

[BE] = Bees

[BU] = Butterflies

[BI] = Birds in general

[H] = Hummingbirds

[N] = Native

PALMS

Palms can have a single trunk or be multi-trunked. Sometimes several single-trunked palms are planted together in a pot by growers, and over time, their roots grow together such that the palm appears to be multi-trunked.

Palm leaves (colloquially known as fronds) are either fan or feather-shaped. Petioles are defined as leaf stems. After the leaves drop, the remaining/persistent petioles are known as boots.

Below are palms for the landscape. Also included are palms you may find around the community, and those are noted as such.

All palms below prefer full sun, ideal for most yards, unless noted otherwise. See the Part 1 Palm Care section for details on caring for your palm.

Palm boots

BISMARK PALM
(Bismarkia nobilis)

Dramatic silvery-blue fan foliage. Makes a statement, but as a large tree, it can overwhelm many homes. Self-pruning. 30'H x 20'W.

CANARY ISLAND DATE PALM
(Phoenix canariensis)

Pretty orange bark and nice feather foliage. Top of trunk resembles a pineapple. Spines near petiole base. Expensive. Susceptible to magnesium deficiency, yellowing, and lethal bronzing, so no longer used as often in landscapes. Slow growing. 40'H x 25'W.

CHINESE FAN PALM
(Livistona chinensis)

Single, but sometimes created as multi-trunked (use for privacy). Fan palm with drooping leaflet tips. Spikey petioles. Black fruits. Self-pruning. 25'H x 12'W.

DWARF PALMETTO
(Sabal minor)

Native palm that feeds wildlife. Rarely forms a trunk. Blue or green foliage. Good for moist, shady areas. Wide spreading, so typically used more in community areas. 8'H x 8'W. Scrub Palmetto (Sabal etonia) is a bit smaller, 5'H x 5'W, and a good alternative for dry, sunny areas. [N]

EUROPEAN FAN PALM
(Chamaerops humilis)

One of the smaller height palms, with fan leaves, and slow growth. It will create pups that make it multi-stemmed, or to control width, cut all or some of the pups off as they appear. Spikey petioles, careful when pruning. Full sun or part shade. Averages 8'H x 12'W (multi-stemmed).

FOXTAIL PALM
(Wodyetia bifurcata)

Beautiful bushy foxtail-like fronds. Trunk is smooth grey with leaf scar rings, and top portion is smooth green. Self-pruning. Will get hit by a freeze, but most recover. 30'H x 15'W.

LADY PALM
(Rhapsis excelsa)

For full or partial shade, a smaller palm with multiple thin stems. Will spread via underground rhizomes - plant in a pot in the ground to help contain it. Can also be grown indoors. Keep it somewhat moist. 8'H x 8'W.

PINDO PALM
(Butia capitate)

Pretty blue-green arching foliage and wide trunk. A great accent specimen plant. Fruit can get messy, but is used to make jelly tasting like pineapple and mango. Slow growing. 15'H x 12'W.

QUEEN PALM
(Syagrus romanzoffiana)

Very popular and used frequently due to its graceful foliage and inexpensive price. Smooth grey bark with leaf-scar rings encircling the trunk. Messy fruits. Can blow over in storms and be prone to fusarium wilt, which causes a quick death. 40'H x 20'W.

RIBBON PALM
(Livistona decora)

Delicate fan leaves with split segments that bend in the middle and droop, looking like ribbons when dancing in the wind. Bark is grey with rings. Spikey petioles facing inward. 30'H x 10'W.

ROEBELLINI PALM
(Phoenix roebellinii)

Also known as the Pygmy Date Palm. Very popular due to its small size and graceful frond foliage. Single stem, but generally grown in clumps of 3 and sometimes 4, that ultimately arch outward away from each other. Slender trunks, fuzzy when young, grey with age. Spined petioles. Will get hit by a freeze and turn brown, but should return. Very slow growing. Multi-trunked 8'H x 8'W.

SABAL PALM
(Sabal palmetto)

Also known as the Cabbage Palm. Native, our Florida state tree, feeds wildlife. Fan palm with criss-cross patterned upright boots or sometimes without boots. Slow growing. 40'H x 10'W. [N]

SAW PALMETTO
(Serenoa repens)

Small height, but wide fan palm in green or silver. Does not form a trunk (unless very old). Fruit is used to help the prostate. Wide spreading, so used mostly in community areas. 6'H x 12'W.

SYLVESTER PALM
(Phoenix sylvestris)

Beautiful feather palm with diamond-patterned bark, golden-orange when new. Large spikes on petioles. Can trim out yellow fruits (females can be messy). As a member of the Date Palm (phoenix) family, susceptible to lethal bronzing. 40'H x 25'W.

WINDMILL
(Trachycarpus fortunei)

Fan palm with distinctive fibrous trunk that looks like burlap. Part shade. 20'H x 7'W.

PALMS FOR INDOORS OR LANAIS

Some palms are sold as houseplants for Florida rooms or lanais. They are exceptionally prone to pests, especially if moved outdoors during the summer and brought back indoors in fall, or if purchased from a garden center where stock is stored outdoors. My experience, after trying many, indicates that the Parlor Palm, also known as the Neanthe Bella palm (Chamaedorea elegans), fares best. It grows 2-3' tall and prefers shade or bright indirect light. It is frequently sold in a small pot and with indoor houseplants.

PALM IN A TREE

This palm was found in South Florida growing naturally inside a Banyan tree! It is included in this book because it is so unusual looking, and if you are traveling in that area, you may question what you are really seeing.

CYCAD "PALMS"

These plants are called palms but are actually in the Cycad family. They are small and slow-growing, and have male and female plants, which vary the fruit appearance. Fruit and seeds are poisonous to pets and humans.

CARDBOARD PALM
(Zamia furfuracea)
Mounding growth habit. Sun or shade. 3-4'H x 5'W.

COONTIE PALM
(Zamia floridana)
Native plant for part shade. 2'H x 2'W. [N]

SAGO PALM
(Cycas revoluta)
Very susceptible to cycad scale. Full sun or part shade. 5'H x 3'W.

Part 3 - Plants

TREES AND LARGE SHRUBS

All of these plants are hardy evergreens unless mentioned otherwise.

ARIZONA CYPRESS
(Cupressus arizonica)

Large pyramidal conifer-looking tree with stunning silver-blue feathery foliage. Full sun. Needs room. 15'H x 8'W. [N]

BALD CYPRESS
(Taxodium distichum)

Native. Similar to a pine, but is deciduous, losing its needles in fall. Tolerates water, grown in or near ponds in community areas. Has interesting "knees", stumpy-like roots that grow above ground. 50'H x 25'W. [N]

BAMBOO
(Bambusa 'Golden Goddess')

'Golden Goddess' is the usual form grown as a landscape plant. It is a clumping bamboo, rather than a running type (stay away from), with beautiful golden stalks when mature. It has a tropical and airy look and is nice for softening up walls. Full sun. It does get large. 8'H x 6'W.

BIRCH, RIVER
(Betula Negra 'Duraheat')

Beautiful white and tan exfoliating bark with age and distinguishable serrated leaves. 'DuraHeat' is the variety grown in Central Florida. Yellow foliage in fall. Deciduous. 25'H x 15'W. [N]

Part 3 - Plants

BOTTLEBRUSH TREE
(Callistemon spp.)

Upright   Weeping

Orange "bottlebrush"-shaped flowers spring-summer. Flowers attract wasps, so locate wisely. Needs adequate water to get established. Full sun. 'Red Cluster' is an upright form, 12'H x 5'W. 'Red Cascade' is a pretty weeping form, 15'H x 6'W. (There is also a dwarf form, discussed under Shrubs.) [BE, BI, BU, H]

CAMELLIA
(Camellia spp.)

Large beautiful single or double flowers in colors ranging from reds, pinks, and whites, with yellow stamens. Two types: Camellia sasanqua is fall-winter blooming and can take some sun. Camellia japonica is winter-spring blooming, needs full shade, and is larger. Acidic. Susceptible to Camellia dieback, which is the sudden death of new twigs, which then need to be removed. Size varies considerably due to the many cultivars, averaging 6-12'H x 5'W. (There are also dwarf forms, discussed under Shrubs.)

CASSIA
(Senna bicapsularis or spp.)

Small tree covered with beautiful yellow flowers atop compound leaves in spring and again in fall. There are some native varieties, and some bush varieties that only bloom in fall. This tree is frequently confused with Senna pendula var. glabrata, which is invasive especially in South Florida. It will most likely not be sold with the Latin name on the label in local nurseries. Attracts butterflies. 10'H x 8'W. [BU]

CHASTE
(Vitex agnus/castus)

Native deciduous tree with beautiful lilac to bluish-purple flowers in spring. A real magnet for butterflies and all kinds of native bees, including the beautiful green sweat bee. Fast growing. Full sun. There are many varieties, some more bush-like than tree-form, but the usual native variety is the one grown. 10-20'H x 10-15'W. [BE, BI, BU, N]

CLEYERA
(Ternstroemia gymnanthera)

Glossy green foliage wih new red growth. Can be a large shrub (perhaps for privacy hedge) or trained into a small tree. 8-10'H x 6'W.

CREPE MYRTLE
(Lagerstoemia spp.)

A true southern plant. Summer flowering, in many colors ranging from reds to pinks to white. Can be single or multi-trunked. Some have beautiful red fall foliage or pretty peeling bark. Deciduous. Do not commit "crepe murder" when pruning. Full sun or part shade. Size varies considerably, from a shrub 6'H x 4'W, all the way up to a tree 25'H x 20'W. [BE, BI, BU]

HOLLY TREE
(Ilex spp.)

Known for their red or orange berries in fall-winter. Most varieties need male and female plants to produce berries on the female. You will not know which you have until fall, so buy then. Attracts birds to eat berries. (After growing all winter, robins migrating north in spring would stop at my holly and devour all the berries in 20 minutes.) Full sun/part shade. Acidic. [BI, N]

7 female hollies, and 1 male

Native Yaupon

Size varies depending on variety, and there are many. For example, 'Dahoon' is a larger native holly used around ponds, 25'H x 10'W. 'Nellie R. Stevens' is 15'H x 8'W, 'East Palaka', a type of American holly, is 35'H x 10'W. 'Sky Pencil" is a columnar form for tight spaces, 8'H x 2'W.

Yaupon hollies (Ilex vomitoria) are native, slow-growing, and have a somewhat scraggly appearance, usually as a weeping form. They produce fruit in the fall that attracts wildlife. 25'H x 10'W. (There is also a dwarf form, discussed under Shrubs.)

HONG KONG ORCHID TREE
(Bauhinia blakeana)

Large pink orchid-like flowers summer to fall and distinctive rounded leaves. It usually needs strong staking and will lose its leaves in wind. 20'H x 20'W.

JAPANESE BLUEBERRY
(Elaeocarpus decipiens)

Red new foliage over shiny green, with inedible dark blue berries that birds love. Usually a pyramidal shaped large bush that can be used in a privacy hedge, or shaped to a small tree. 15'H x 10'W. [BI]

ITALIAN CYPRESS
(Cupressus sempervirens)

Tall, very columnar shape. Use in homes with tall roofline, otherwise they will dwarf a house. Fast growing. Full sun. 30'H x 4'W.

LIGUSTRUM TREE
(Ligustrum japonicum)

Viburnums have mostly replaced this tree-form of Ligustrum, which is very prone to fungal black spot disease. Usually kept pruned at the base to keep foliage high above the soil as a preventative measure. 10'H x 6'W.

MAGNOLIA
(Magnolia grandiflora)

Southern Magnolias have glossy foliage with rust color on leaf undersides, and large white saucer flowers in spring through summer. Leaves drop and get messy. Full sun. Acidic. Larger varieties are found growing in woods, but smaller varieties such as 'Little Gem' are 25'H x 10'W, and are grown in landscapes. [N]

MAPLE, FLORIDA RED
(Acer rubrum)

Native maple that tolerates moisture and can be found around ponds. Two common varieties: 'Summer Red' has red new foliage and branches, and turns yellow in fall. 'Flame' has pretty red, orange, and yellow foliage in fall. Full sun. Not as wide as northern varieties for creating shade, but rather has a more cylindrical shape when young that widens slightly with age. Deciduous. Birds love to perch in this tree. 35'H x 15'W. [BI, N]

'Florida Flame'

'Summer Red'

NORFOLK ISLAND PINE
(Araucaria heterophylla)

Sold during the holidays in cute little red containers as a miniature Christmas tree. Not a true pine. Has distinctive layering with age. Grows 60' tall, and although slow growing, it is not recommended for homeowner yards due to its height. Can be used as a houseplant instead, given enough humidity and bright light. If you must try it in your landscape, cut the central leader/trunk when it reaches your desired height.

OAK
(Quercis spp.)

Many different varieties. Native live oak (Quercus virginiana) is 60-80'H x 60-120'W, depending on variety. The second largest in the state of Florida is located at Lake Griffin State Park in Fruitland Park. Laurel oak (Quercus laurifolia) is most frequently used in landscapes and communities due to its smaller size of 50'H x 35'W. However, it is also less hurricane-proof (photo from Hurricane Milton). Birds love to hang out in oak trees. [BI, N]

Live Oak

Laurel Oak

Laurel Oak after storm

PINE

(Pinus elliottii var. densa)

Unlike northern areas of the country, even "dwarf" varieties get tall in Central Florida. The smallest one, the native Slash Pine 'Densa', is the one most likely to be found in landscapes, at 35'H x 25'W. [BI, N]

PODOCARPUS

(Podocarpus macrophyllus 'Maki')

'Maki'.   'Icee Blue'

Tall, narrow, cylindrical form with soft green foliage and needles that look similar to Bottlebrush. Used as a hedging plant. Slow growing. Full sun or part shade. 25'H x 6'W.

'Icee Blue' is a pyramidal variety. It has beautiful blue foliage, offset by new lighter foliage that seems to glow. Full sun. 20'H x 6'W.

POWDERPUFF
(Caliandra haematocephala)

Large shrub with a tree-like vase shape with unusual pink flowers from fall to spring. Full sun/part shade. Other varieties are invasive. 8'H x 8'W.

PURPLE LEAF PLUM
(Prunus cerasifera 'Krauter Vesuvius')

Small tree with purple foliage and pink flowers in spring. Ornamental; does not produce fruit. Full sun. Deciduous. 25'H x 15'W.

TEA OLIVE
(Osmanthus fragrans)

Glossy foliage with extremely fragrant white flowers in spring. Full sun. Prone to tea scale and black sooty mold. 8-10'H x 6'W.

TRUMPET TREE
(Tabebuia spp. or Handroanthus spp.)

Deciduous tree with yellow or pink flowers in spring that appear before the foliage. A showstopper from the road. Fuzzy seed pods over winter. Full sun. 25'H x 25'W. [Note: Many in the Tabebuia family have been renamed to Handroanthus.]

VIBURNUM
(Viburnum odoratissimum or suspensum)

Sweet Vibrurnum   Ed showing height

Sweet Viburnum (Viburnum odoratissimum) has shiny foliage, red new growth. Tiny white flowers in spring. Used primarily as a hedge plant. Full sun. Fast growing, gets tall. Keep pruned to prevent lower legginess. 20'H x 6'W. A smaller version is Sandankwa Viburnum (Viburnum suspensum), 6'H x 4'W. [Bl, N]

Sandankwa Viburnum

SHRUBS, HARDY

All of these plants will survive a freeze and will not result in damage. They are small to medium in size.

ABELIA
(Abelia x grandiflora)

White flowers summer-fall. Sun/part shade. May drop its leaves. Many beautiful varieties with varying height. 'Miss Lemon' has yellow foliage and is 3-4'H x 3-4'W. 'Radiance' also has yellow foliage and is 3'H x 3'W. 'Kaleidoscope' has yellow spring foliage that turns to orange in fall and is smaller at 2½'H x 2½'W.

ANISE
(Illicium parviflorum)

Anise 'Florida Sunshine' is a native bush with yellow to bright lime colored foliage, scented when crushed. (Other Florida varieties also exist.) Part shade. Fast growing. 6-8'H x 4-6'W. [N]

AZALEA
(Rhododendron spp.)

Beautiful flowers, ranging in color from reds to pinks to whites, blooms in spring, with some varieties reblooming. Part shade. Acidic. Some common Azaleas for our area:

- **Encore** azaleas can take more sun. They bloom 3 times per year. Can be more expensive. Size ranges based on variety, but generally 3-5'H x 3-4'W.

- **Red/Pink Ruffles** blooms 2-3 times per year. 2-3'H x 2-3'W.

- **Formosa 'Southern Charm'**, pink in spring, 6-8'H x 6-8'W.
- **Native** azaleas are deciduous, fragrant, have an open habit, and can be found in the woods. 6'H x 6'W.

BEAUTYBERRY
(Forestiera segregata)

Not commonly found as a landscape plant, but mentioned here because of its value to wildlife. Grown for its beautiful purple berries, which also attract birds. Full sun/part shade. Flowers on new wood, so pruning will increase berries. Drought tolerant. Repels mosquitoes. 6-8'H. [BI, N]

BOTTLEBRUSH, DWARF
(Callistemon viminalis 'Little John')

'Little John' is a dwarf version with blue-green foliage and orange-red flowers, smaller than the tree varieties, but still attracts hornets, so be careful of placing it near the front entry. 2-3'H x 2-3'W. [BE, BI, BU, H]

BUTTERFLY BUSH, DWARF
(Buddleia hybrids)

Beautiful flower clusters in various shades of purple, pink, or white that attract butterflies. Keep it pruned to keep it more full. Susceptible to pests such as mites and soil nematodes. Full sun. Dwarf hybrids 3'H x 3'W. [BU]

CAMELLIA, DWARF
(Camellia spp.)

Beautiful single or double flowers in colors ranging from reds, pinks, and whites, with yellow stamens. Acidic. Dwarf varieties of Camellia sasanqua are perfect landscape plants, approximately 3'H x 3'W.

FOXTAIL FERN
(Asparagus densiflorus 'Myers')

Tropical looking, lime-green unusual foliage like a fox's tail. Great under palms. Part shade. 2'H x 4'W.

GARDENIA
(Gardenia jasminoides)

Very fragrant white flowers spring through summer. Full sun/part shade. Acidic, but you may need to add iron also to keep green. Many varieties, and although some are larger, they are generally 3-4'H x 3-4'W.

HOLLY, SHILLINGS
(Ilex vomitoria 'Schilling's Dwarf')

Native, very dwarf form of Yaupon Holly. Small white flowers in spring. 2-3'H x 2-3'W. [N]

HOLLY, BUFORD DWARF
(Ilex cornuta 'Bufordii Nana')

Buford is a compact variety, frequently used as a hedge, or grouped in a foundation planting. 5-6'H x 6'W. [BI]

HYDRANGEA, OAKLEAF
(Hydrangea quercifolia)

Hydrangeas do not always grow successfully in Central Florida. Oakleaf is native, however. Cone shaped clusters of white flowers turn to pink as they age. Leaves are oak-shaped, and red in fall before they drop. Acidic. Part to full shade. Normally 6-10'H x 6-10'W, but 'Pee Wee' dwarf Oakleaf grows 4'H x 4'W. [BU, N]

Nini's Oakleaf Hydrangea

INDIAN HAWTHORNE
(Rhaphiolepis indica)

Nice rounded shape, with red new foliage and small fragrant white flowers in spring. Normally planted in groups. Full sun/part shade. Acidic. Slow growing. Becomes woody in the interior, so keep pruned if a smaller size is desired. 4'H x 4'W.

Practical Landscaping

JUNIPER
(Juniperus 'Parsonii')

One of the biggest attributes of Juniper is its drought tolerance, particularly for difficult south or west full-sun areas. Many varieties with foliage ranging from green to blue to yellow, with blue berries. Dwarf varieties spread to form a wide ground cover, usually green and 1-2'H x 6'W. Taller columnar varieties also exist for use in landscapes, typically blue and 8-10'H x 4'W.

LIGUSTRUM, DWARF
(Ligustrum spp.)

Bush forms of Ligustrum vary in color and size. Full sun/part shade.

- **'Jack Frost'** (Ligustrum japonicum 'Jack Frost') has creamy-edged leaves. Frequently used as a foundation plant, it will need pruning to keep it small. 8'H x 6'W.

- **'Howardi'** (Ligustrum japonicum 'Howardi') has variegated yellow and green leaves. 6'H x 6'W.

- **'Sunshine'** (Ligustrum sinense 'Sunshine') has bright tiny yellow leaves that shine from the road. 4'H x 3'W.

Part 3 - Plants

LOROPETALUM
(Loropetalum chinense)

Beautiful purple foliage with pink tassle-like flowers in winter-spring. They get leggy, so keep on top of pruning. Acidic. Many new cultivars are being developed, watch for size. Some, like 'Purple Pixie', are dwarf bushes only 1-2'H x 4'W, others are larger dwarfs, and yet others are trees.

'Jazz Hands' is an unusual variety with new growth of pink, white, and purple above mature purple leaves. Color is more vivid in heat and partial shade. 5'H x 4'W.

'Jazz Hands'

MEXICAN PETUNIA
(Ruellia brittoniana 'Purple Showers')

Purple (or pink or white) flowers on grass-like foliage summer-fall. Ruellia simplex is invasive, be sure to choose sterile varieties such as the 'Mayan' series or 'Purple Showers'. Attracts butterflies and rabbits (may need to use rabbit spray). Sun/part shade. 2-3'H x 2-3'W. [BU]

NANDINA
(Nandina domestica 'Fire Power')

There are different varieties of Nandina. Be careful of those that are invasive. 'Firepower' is a smaller variety with lime green foliage in spring, then striking red, orange, yellow, and purple fall-winter. Full sun/part shade. 2-3'H x 2-3'W.

OLEANDER
(Nerium oldeander)

Vivid flowers in shades of pink in summer. Dwarf varieties 3-4' tall,

regular 10-18' tall. Full sun (for best flowering) or part shade. All parts are very poisonous.

Susceptible to the oleander caterpillar (non-stinging), which will skeletonize leaves, then defoliate within 2 weeks, and could kill the plant if it happens yearly. For those reasons, although pretty, I do not recommend growing this plant.

PITTOSPORUM
(Pittosporum tobira 'Variegatum')

Most often, the variegated version is planted, however, branches may suddenly appear in the original green form (prune them out). Fast growing and can get tall, so keep on top of pruning. 6-8'H x 6-8'W.

PODOCARPUS, DWARF
(Podocarpus macrophyllus 'Pringles')

Dwarf variety 'Pringles' has soft green foliage, similar to Bottlebrush. Excellent plant for hiding utilities or can be used as a low hedge. Slow growing. 5'H x 3'W in full sun or 2-3'H x 2'W in part shade.

ROSE
(Rosa spp.)

Beautiful flowers with colors of red, pink, peach, yellow, and white. They bloom profusely in spring, and many varieties, including the ones listed here, bloom sporadically year-round. Full sun. Use special rose fertilizer. Prune off old flowers if you wish to extend blooming periods. Leaf/black spot is common, caused by rain and overhead watering; use anti-fungal spray.

- **Knockout** varieties have beautiful colors, such as fuscia and yellow. Can get chili thrips. 3-4'H x 3-4'W.

Marti & Ken's Knockout roses

- **Drift** roses come in red, pink, peach, and yellow. My favorite is 'Apricot', with peach flowers and touches of pale orange, and yellow centers. 'Coral' is used frequently for its bright colors that go with many landscape themes. Smaller than Knockout, 1-2'H x 1-2'W.

'Apricot'

'Coral'

- **Grafted** varieties grown on Fortuniana rootstock, which has adapted to Florida's tough climate, such as Nelson's Florida Roses.

TEXAS SAGE
(Leucophyllum frutescens)

Native to Texas as the name implies. Very pretty blue-silver foliage (although green varieties also exist). Small pink flowers spring-fall that bloom in high humidity or after rains. Full sun, likes it hot and dry. Doesn't like fertilizer or compost. Pyramidal shape when small, becoming rounded with age. 5'H x 3'W. [N]

WALTER'S VIBURNUM
(Viburnum obovatum)

Native shrub with new red foliage and small white flowers in late winter to spring. Drought tolerant. Spreads by runners that are tough to remove. Dwarf varieties are 3-4'H x 3-4'W. [BI, N]

SHRUBS, TROPICAL

These shrubs are considered tropical, and foliage will turn brown and be damaged by a frost. But most are root-hardy and will come back in the spring. Tropical plants are used primarily because they typically provide more color than hardy plants.

ALLAMANDA
(Allamanda neriifolia)

Beautiful showy yellow trumpet flowers summer-fall. Full sun/part shade. Milky sap may irritate skin, so wear gloves for pruning, and all parts of the plant are poisonous if ingested. If yellowing out during rainy season, add iron. Two types of shrubs: Bush is 5'H x 4'W, Compact Bush is 2'H x 2'W and with smaller flowers. (There is also a vine form, see Vines section.)

Part 3 - Plants

BIRD OF PARADISE
(Strelitzia reginae)

Interesting orange and blue bird-like flower on very large upright tropical looking leaves (like a banana tree, but with skinnier leaves). There is also a white variety, which is larger. Not to be confused with False Bird of Paradise. Fertilize regularly. Sun (may keep it shorter) to part shade. 5-6'H x 3'W.

BUSH DAISY
(Euryops pectinatus)

Bright yellow daisy-like flowers summer-fall. Keep pruned for compactness. Needs excellent drainage. Full sun. 2'H x 2'W.

BUTTERCUP BUSH
(Turnera ulmifolia)

Showy yellow flowers that look like large buttercups, on glossy green foliage. Blooms in summer and potentially year-round. Likes moisture. Attracts butterflies. 3'H x 2'W. [BU]

Buttercup flower with green Lynx spider

COPPER LEAF
(Acalypha wilkesiana)

Mottled copper colored leaves. Sun/part-shade. This plant is Zone 10b, so it will receive damage during a frost. A large bush at 8'H x 3'W.

CROTON
(Codiaeum variegatum)

Colorful pretty red, orange, yellow, and green leaves. Sun enhances color, but too much causes fading. Irritating sap. Some varieties are not as hardy in our area, but 'Petra' is the hardiest, with 'Mammy' not far behind. Full sun or part shade. Sometimes recommended as a houseplant, but it must receive enough sun to maintain color. 3'H x 2'W.

'Petra'

'Mammy'

CROWN OF THORNS
(Euphorbia milii)

A shrubby succulent plant known religiously for its thorns. Colorful bracts in red, rose, peach, or yellow that surround small yellow-flowered centers. Green or variegated yellow foliage. 2'H x 2'W.

DEWDROP
(Duranta repens or erecta)

'Gold Mound'　　'Purple Showers'

Dewdrop has yellow foliage and pretty ½" golden fruit, poisonous to humans, but may attract birds. Spring-fall blooming. Sun/part shade. Prone to whitefly. 'Gold Mound' is a smaller plant at 2'H x 2'W. 'Purple Showers' has violet flowers with pretty gold fruit, and is frequently sold as a standard.

DIPLADENIA
(Dipladenia splendens)

Related to Mandevilla, but a bush form, with smooth finely pointed glossy leaves on stems that grow down and hang, rather than the upward vine form and textured wider leaves of Mandevilla. Showy trumpet flowers, usually red, are smaller than on Mandevilla. A zone 10 plant, it will get hit in frost, and may not return. 2'H x 2'W.

ESPERANZA
(Tacoma stans 'Sun Trumpets Yellow')

There are different varieties and sizes, with 'Sun Tumpets Yellow' being a dwarf bush. This and larger versions can be pruned into a small multi-trunked tree. Showy yellow trumpet flowers in summer. Seed pods in fall. Drought tolerant. 6'H x 6'W.

FALSE BIRD OF PARADISE
(Heliconia dune)

Orange bracts that look like flowers, with leaves that resemble Bird of Paradise, spring-summer. Other colors exist, but orange is the most popular. Sun to part shade. 2'H x 2'W.

FIREBUSH
(Hamelia patens)

Lovely orange flowers spring-summer. Dwarf varieties, but also much larger ones. 'Lime Sizzler' has yellow variegated foliage. Sun/part shade. 4'H x 4'W. [BU, H, N]

FIRECRACKER FERN
(Russelia equisetiformis)

Orange-red flowers almost year-round on ferny arching foliage. Landscapers mutilate this plant by pruning off its beautiful arching form and creating mounds. Will spread via suckers that need to be controlled, or keep in a pot. Attracts hummingbirds, even when in a pot on a patio. Full sun/part shade. 3'H x 4'W. Firecracker Plant (Russelia sarmentosa) is similar, but leaves are oval instead of fern-like, and the plant is smaller in width, hardier, and can take more shade. [BU, H]

GINGER, VARIEGATED/SHELL
(Alpinia zerumbet 'Variegata')

Tropical looking with large upright yellow and green variegated leaves. Part shade. 3'H x 3'W.

HIBISCUS
(Hibiscus rosa-sinensis)

This plant screams tropical more so than any other. Big gorgeous blooms in various colors such as red, pink, orange, yellow, and multi-colored. Blooms spring-fall. Fertilize often for blooms. Needs good drainage. Its drawback is being prone to many problems, including mealybugs, hibiscus sawfly, yellowing leaves, bud drop (moisture or a midge), etc. Keep up with pest management. This plant is Zone 10, and will turn brown in a freeze, although most return. Sun. Averages 5-6'H x 3-4'W.

IXORA
(Ixora coccinea)

Flowers in red-orange (hardiest) and golden yellow. Blooms spring-fall. Acidic (foliage yellows in alkaline). Full sun/part shade. Two varieties: 'Taiwanese' is better sized for foundation plantings and smaller landscape areas at 2-3'H x 2-3'W. 'Maui' is a larger bush at 4-5'H x 4-5'W. [BU]

'Taiwanese' 'Maui'

JATROPHA
(Jatropha integerrima 'Compacta')

The dwarf bush version of this tree has pretty red flowers most of the year. The non-dwarf version is a small tree. If no tag, and space in your yard is a concern, be sure to ask if the plant you are purchasing is indeed dwarf. Many nurseries do not carry dwarf. Milky sap can irritate skin. Attracts birds and butterflies. Full sun. 5'H x 4'W.

MEXICAN HEATHER
(Cuphea hyssopifolia)

Cat's Mexican Heather

Small plant with purple flowers atop heather-like foliage. May be short-lived. Prune to keep compact. Full sun or part shade. 1½'H x 1½'W.

MUSSAENDA
(Mussaenda glabra)

Interesting deciduous shrub with small yellow flowers with striking white petals on top of green leaves. Blooms in spring through summer. Does best with regular irrigation or moist soil. Part shade. 6'H x 4'W. [BU]

OYSTER PLANT
(Rhoeo spathacea)

Purple *Yellow*

Low-growing purple green variegated upright foliage plant. Also, unusual yellow variety exists. Overwatering causes disease. Sun/part shade/shade. Can spread. 1'H x 1'W.

PHILODENDRON
(Philodendron x 'Xanadu')

Large   Dwarf 'Xanadu'

It has a tropical, almost Jurassic Park, feeling with large leaves with cut foliage, and does best in part shade. 'Xanadu' is a smaller philodendron, 2-4'H x 4-6'W. The larger variety is extremely large and seldom usable in landscapes.

PLUMBAGO
(Plumbago auriculata)

Beautiful blue flowers (like phlox) almost year-round. Will put out suckers, but not obnoxiously. Full sun. Multiple species of butterflies enjoy this plant. The variety 'Imperial Blue' stays a bit smaller than older varieties at 3-4'H x 3'W. [BU]

SERISSA
(Serissa foetida)

Creamy variegated trim on small leaves, pink buds that turn into white flowers spring-summer. Keep pruned for fullness. Used in bonsai also. Full sun/part shade. 2-4'H x 2-4'W.

SHEFFLERA/TRINETTE
(Shefflera arboricola 'Trinette')

'Trinette' is a shrub with variegated yellow and green palmate foliage. Prune out wayward branches and to create a denser rounded shrub. Keep on the dry side. You may have to keep rabbits at bay. Sun/part shade. 4'H x 4'W.

SHRIMP PLANT
(Justica brandegeana)

Shrimp plant has flowers that resemble shrimp scales. The red version is hardier than the yellow Golden Shrimp plant (Pachystachys lutea). Summer blooming. Sun/part shade. 3'H x 3'W. [H-red varieties]

THYRALLIS
(Galphimia glauca)

Many small yellow blooms almost year round. Full sun. 6'H x 4'W.

TI PLANT
(Cordyline fruticosa)

Pretty burgundy & pink striped foliage. As with all Cordylines, it will lose lower foliage as it grows. Full sun (more color) or part shade. As a Zone 10 plant, this will turn brown with a freeze. 4-5'H x 2'W.

TIBOUCHINA
(Tibouschina urvilleana)

Very pretty purple flowers almost year-round on velvety green-blue leaves with white stripes. Compact form is 3'H x 2'W. Sometimes they are sold as tree-like standards.

YESTERDAY, TODAY, TOMORROW
(Brunsfelsia grandiflora)

Flowers start out as deep violet, then turn to light violet, and then white as they age over 3 days. Blooms spring-fall. Berries are poisonous to pets. Full sun/part shade. Large shrub at 8'H x 6'W.

PERENNIAL FLOWERS

Perennial flowers remain year after year (unlike annual flowers discussed in the next section). Although perennials are generally not used as extensively in the landscape compared to shrubs, below are the most common ones. Others are found in community areas. All take full sun unless otherwise noted.

AFRICAN IRIS
(Dietes vegeta)

White flowers with yellow and purple streaks on grass-like foliage spring-summer. Full sun/part shade. They like moisture but also tolerate dryness. Used mostly in community areas and medians.

AGAPANTHUS
(Agapanthus africanus)

Also called Lily of the Nile. Blue, purple, or white flowers bloom in early summer on grass-like foliage. Blue flowers may revert to white. They like to be crowded in groups, and need plenty of fertilizer to flower. Full sun/part shade.

ANGELONIA
(Angelonia angustifolia)

Small purple or pink flowers on upright stems. Sometimes treated as an annual, this will get hit by frost, but usually comes back. Drought tolerant.

BLACKBERRY LILY
(Belamcanda chinensis)

A grass-like plant topped with yellowish-orange flowers with red speckles and stamens in summer.

BLANKET FLOWER
(Gaillardia x Grandiflora)

Brightly colored daisy-like multi-colored flowers of red, orange, and yellow with dark centers. Blooms spring through fall, and will reseed to act as a perennial. [BU, N]

BLUE DAZE / BLUE MY MIND
(Evolvulus glomeratus)

Very popular low-growing plant with small blue flowers and soft blue-green leaves. Two varieties that look similar – the species, which has slightly larger flowers, and 'Blue My Mind,' with smaller flowers but more of them. Generally used en masse or as a border. Blooms spring-fall, but almost year-round. Flowers close in the afternoon. Sun/part shade. 1'H x 2'W.

BULBINE
(Bulbine frutescens)

A grass-like plant, the variety 'Hallmark" is topped with small orange and yellow flowers in summer. Because it is not a showy plant, it is planted in masses mostly in community areas. Be sure to have good drainage. Sun/part shade.

CANNA LILY
(Canna x generalis)

Meant for our summer heat, Canna Lilies are large plants with leaves that are large and very tropical looking. Breeding has created interesting variegated foliage patterns beyond green. Showy large blooms include colors of red, orange, peach, yellow, and even multi-colored. They need adequate moisture and fertilizer. Full sun/part shade. Most brown out or die back in winter. 3-4'H x 2'W.

CHRYSANTHEMUM
(Chyrsantemum spp.)

We do have mum season in Central Florida! Mums here are hardy compared to up north where most are considered annuals. They come in many colors, including yellow, pink, orange, and more. They are sold in fall, but bloom earlier also and through winter.

CONEFLOWER
(Echinacea spp.)

The native coneflower is pink, however, there are other stunningly beautiful colors nowadays in red, orange, and yellow. Drought tolerant. [BE, BU, N]

COREOPSIS
(Coreopsis spp.)

Our Florida state wildflower. A daisy-like but smaller flower, typically yellow, although other colors exist. Native varieties attract butterflies. Short-lived perennial, 3-4 years. Drought tolerant. [BU, N]

CROCOSMIA
(Crocosmia spp.)

Beautiful orange-red tubular flowers with yellow stamens atop upright grass-like foliage in summer that attract hummingbirds. 2'H x 2'W.

CROSSANDRA
(Crossandra infundibuliformis)

'Orange Marmelade' has pretty creamsicle-colored orange flowers in summer on shiny foliage. Acidic. This is a Zone 10 plant, so it may not return after a freeze. Part shade. 1½'H x 1½'W.

DAHLIA
(Dahlia spp.)

Beautiful large flowers in many different colors and sizes in spring through early summer. In northern areas of the country, Dahlia bulbs need to be dug up every year and replanted in the spring, but not here in Central Florida. However, many will not tolerate our hot summers, so seek out heat tolerant varieties. Best in part shade. Susceptible to many pests and diseases.

DAYLILY
(Hemerocalis spp.)

Daylilies grow in northern areas of the country, as well as in Central Florida. Their grass-like foliage adds interest to the landscape. Bloom occurs in spring, although many varieties are re-blooming, which means they will bloom into summer as well. Various bloom colors including yellows, oranges, and reds. Common reblooming varieties include 'Stella d'Oro' (golden yellow), 'Happy Returns' (yellow), 'Yangtze' (lemon yellow), and 'Little Business' (red). Full sun/part shade. Average 2'H x 2'W.

DIANTHUS
(Dianthus spp.)

Also called Pinks. Cold-hardy varieties up north are perennials, although we also had tropical annual versions there as well. In Central Florida, however, Dianthus are those tropical versions, and they are usually categorized as a cool weather annual. However, depending on variety, they usually will live through a freeze. Part shade will get them through our heat. Flower colors are typically multi-colored in the red, pink, and white range, with varieties having an eye in a different color. Sizes vary from low-growing mounds 12"H x 12"W, such as my perennial Supra Red, to taller versions.

GAURA
(Gaura spp.)

Another plant from up north that does well in Florida as well. Known for its drought tolerance. 'Crimson Butterflies' is a popular variety with reddish-pink flowers floating atop red grass-like foliage, giving an airy appearance. Summer blooming.

GERBERA DAISY
(Gerbera jamesonii)

Standout daisy-like colorful flowers in bright colors of red, pink, orange, and yellow. Blooms early summer through fall and sometimes year-round. Plant slightly above ground level to prevent crown rot. Sun/part shade.

LANTANA
(Lantana spp.)

Yellow, orange, or multicolored flowers spring-fall. Make sure to buy non-invasive varieties! Costa Farms, Bandana series, or natives are okay. May suffer in a freeze. Cut back in spring if woody. Size varies, but generally 2'H x 3'W. [BU]

Invasive — *Watch the tags*

LIVINGSTONE DAISY
(Dorotheanthus Bellidiformis)

A variety called Mezoo is a low-growing succulent plant (6" tall) with small red-pink daisy-like flowers that contrast beautifully against the thick variegated green and white leaves. Flowers close at night. Drought tolerant.

LYSMACHIA
(Lysmachia spp.)

Creeping Jenny   Waikiki Sunset

Low growing plant for part shade. 'Goldilocks' (Lysmachia numularia 'Goldilocks'), commonly known as Creeping Jenny, is a variety used for its rounded golden leaves that form a low mound or drape nicely over a container. 'Waikiki Sunset' (Lysimachia congestiflora) has variegated yellow and green foliage with yellow flowers in spring through summer.

MILKWEED
(Asclepias spp.)

Monarch butterflies are being decimated. They breed only on Milkweed. Use the non-native variety (Asclepias tuberosa), since tropical varieties bloom year-round and may alter normal migration, resulting in butterfly viruses. Native milkweed has blooms in late summer-fall (usually orange). Place in the back of the garden where they can be devoured. If you've inherited a property with non-native milkweed already on it, cut it back to the ground in the fall. Full sun. [BU, N variety]

MONA LAVENDER
(Plectranthus 'Mona Lavender')

Not actually a lavender, but a member of the Plectranthus family. Purple tubular flowers above foliage that is dark green with undertones of purple, especially on the undersides. Blooms late summer through fall. Part shade.

PENTA
(Pentas lanceolata)

Star-shaped tubular flowers in white, red, pink, purple. Attract butterflies. Blooms best in summer, but also blooms almost year-round. Do not overwater. Prune to keep bushy. May be short-lived. Full sun/part shade. Size varies, 1½-3'H x1- 2'W. [BU]

RUDBECKIA
(Rudbeckia fulgida or hirta)

Otherwise known as Black-eyed Susan, it is an annual up north. But in Central Florida, it is considered a perennial. Drought tolerant. [BU, N]

SALVIA
(Salvia spp.)

Many varieties and sizes, some get large. Usually blue, purple, or red flowers all summer long. They attract many beneficials, with red attracting hummingbirds, and purple attracting bees. Full hot sun and drought tolerant. [BE non-red, BU, H (red)]

SOCIETY GARLIC
(Tulbaghia violacea)

A grass-like plant topped with small blue flowers that bloom spring through summer. Since it is not a showy plant, it is more often planted in masses in community areas. It does smell like garlic and some people find it offensive.

STOKE'S ASTER
(Stokesia laevis)

lso grown up north, I have tried varieties from there, and they did not survive in Central Florida. However, the native version did and reseeds. It should be used more in landscapes because of its beautiful large blue flower in summer. [N]

VERBENA

Grow in containers or in your garden. Pretty colors include red, pink, purple, or white. Full sun, drought tolerant. Some varieties are supposedly cold- hardy, but I personally have not found that to be true.

VINCA
(Vinca spp.)

Vinca typically will survive a freeze and live for several years in Central Florida. Blooms almost year-round and in popular colors, including fuscia and red. There are two different types, one with larger flowers and leaves, the other with smaller flowers, but more of them.

Part 3 - Plants

ANNUAL FLOWERS

Annual flowers last one season and they have a preference for either our cool season or warm season. Annuals are typically not used in the landscape unless in containers.

In my community they are used in medians and entrance areas, and they are changed out seasonally.

Although you may occasionally find others, featured below are the most common ones for our Central Florida area.

Median in fall

Median in winter

Median in spring

Median in summer

COOL SEASON ANNUALS

ALLYSUM

Small white or purple fragrant flowers over mounding foliage in winter to spring. Sun/part shade.

DELPHINIUM

Striking deep blue flowers on upright ferny foliage in late winter. Although plant tags will usually designate this as a perennial, it never makes it.

GERANIUM

Same as the Geraniums grown up north, but in Central Florida, they are considered a cool weather rather than warm weather annual. Colorful flowers in bright red, orange, or pink atop green or variegated foliage. To keep them looking their best, deadhead by removing the old flowers.

MILLION BELLS

Colorful blooms, similar to Petunias but smaller, January through June. Many colors, including multi-colored. Spills over a container edge nicely. Sun/part shade.

NEMESIA

Nemsia and Diascia are in the same family as S napdragons, but generally lower growing. Flowers are in the pink, orange, yellow range. Full sun to part shade.

PANSY

Pansies are known for their cold hardiness. They come in all sorts of colors. They will last until the heat of summer arrives.

PETUNIA

Petunias were grown as annuals up north, but in Central Florida, they cannot tolerate our heat and thus are considered cool season annuals. Many colors, and unlike old-fashioned varieties, most are self-cleaning these days.

SNAPDRAGON

Varieties vary in color from red to orange and mixed with yellow or pink. Full sun. Very cold hardy, and sometimes will even make it through summer. Heights vary.

WARM SEASON ANNUALS

BEGONIA

Reiger

Rex

There are several kinds of begonias, including Wax, Reiger, and Rex. Wax begonias have orange, pink, or white flowers with yellow stamens and green or ruddy foliage. They do well in sun or shade and act more as a perennial, blooming continually throughout the year. Reiger begonias come in beautiful bright colors of red, orange, yellow, and pink, but the foliage dies back in cold. Do not water afterward until the foliage starts to regrow in the spring. Rex begonias are known for colorful foliage, including purples and silvers.

Wax

COLEUS

There are two types of Coleus – Sun Coleus and the old-fashioned kind for shade. Coleus is grown for its colorful foliage, in red, pink, orange, yellow, or lime green. Beautiful in containers with matching flowers. Needs moisture.

IMPATIENS

Shade Impatiens　　　　　　　*Sun Impatiens*

There are two types of Impatiens – New Guinea, meant for sun and with elongated leaves, and the original, meant for shade and with smaller rounded foliage. Nice shades of red, pink, orange, and white. The 'Sunpatiens' series includes a beautiful version with variegated green and yellow leaves and peach or pink flowers.

LOBELIA

Small pretty blue or purple flowers in a mass that looks beautiful tumbling over a container or hanging basket. Spring through early summer. Part shade.

MARIGOLD

African

French

Two varieties – the smaller French Marigolds, and the larger African Marigolds, which also take more heat. Marigolds come in oranges and yellows and mixed colors. Prone to snails. If grown organically, French Marigold petals are pretty in your salad!

PORTULACA

Sometimes called Moss Rose or Purslane, this is a low-growing succulent for hot, sunny areas, and it is drought tolerant. It has pretty flowers in pink, red, orange, or yellow shades in summer. Flowers close at night.

SCAEVOLA

Low-growing with purple fan-like flowers with small yellow and white centers. Heat and drought tolerant. Full sun to part shade.

Part 3 - Plants

ZINNIA

Easily seeded. Different varieties range in size and colors from pink, red, orange, and yellow. Prone to insects.

BULBS

The spring-blooming bulbs that you were familiar with up north, like tulips and crocuses, will not grow here (despite the fact that you may find the bulbs for planting in fall at big box stores).

Bulb display

Daffodil in Central Florida

At one time Daffodils were grown in Central Florida, with the small cupped ones handling heat best. But alas, after trying many varieties, most did not come back at all for me, and only a couple lasted a couple of years. Paperwhites, sold around the holidays, are more tropical and can be planted outside, but even those do not typically come back the following year.

We do not get the number of cold days that many bulbs need to bloom (unless you want to do the refrigerator trick for 2 months). The bulbs we can grow in Central Florida are included in the Perennial Flowers section.

Part 3 - Plants

SUCCULENTS

Succulents cover a wide variety of species. Smaller tropical succulents are used in container arrangements. Landscape succulents require full sun and are drought-tolerant. Their spikey appearances add dramatic interest to the landscape. Larger varieties are used as accent plants.

AGAVE
(Agave spp.)

Large stiff green or blue succulent leaves sometimes variegated with yellow. Sharp tips. The mother plant dies after blooming every 10 years, but spawns pups. These should be used as an accent plant in the landscape. They get tall - some are 6' tall, others get as tall as your house.

PRICKLY PEAR
(Opuntia humifusa)

Occasionally someone will want to incorporate a Midwest theme into their yard. Succulents exist for that also, such as the Prickly Pear cactus. Cacti usually prefer dry conditions, so be sure to have exceptional drainage due to our Floridian rainy season.

Practical Landscaping

SEDUM
(Sedum rupestre)

Stonecrop, such as the variety 'Florida Friendly Gold,' is a low-growing yellow ground cover. It has short, thin, stubby leaves with yellow flowers in summer. The stunning yellow color provides a beautiful pop to the front of the garden, even from afar. Drought tolerant. 6"H x 1'W.

SUCCULENTS, TROPICAL

You will find small tropical succulents at garden centers, but they are usually meant for containers in part shade. They include Echeverias, Cacti, Haworthias, and more.

YUCCA, DWARF
(Yucca filamentosa 'Colorguard')

Although there are many varieties of yucca, 'Colorguard' is the one used most often for landscapes. It is a smaller landscape succulent that has beautiful variegated yellow and blue-green sword-like foliage. White flowers occur in later summer on tall stalks. Be careful – foliage tips are sharp. 2'H (without stalk) x 2'W.

YUCCA, SPINELESS
(Yucca elephantipes)

A really tall, interesting yucca that looks almost like a unique palm tree. New pups will form at the base, giving it a multi-stemmed appearance. It will need pruning to reduce height every few years (or you can let it grow to 20'), but the removed stalks propagate easily to make new plants. Be careful of the sharp foliage tips. Can also be grown as a houseplant that reaches only 5' tall.

GRASSES

Grasses add a different form to the landscape than the usual rounded plants, which in turn adds interest. The smaller grasses look good lining a border, whereas the larger ones can be used as privacy plants, to hide utilities, or to soften fences and walls. The larger ones may also harbor rats and, thus, snakes. Unless otherwise noted, grasses need full sun. Featured below are common grasses used in the landscape and community areas.

AZTEC GRASS
(Liriope muscari 'Aztec')

A type of Liriope meant for full sun/part shade. Variegated creamy white and green foliage with white flowers in summer. Great for borders or along walkways. 1-2'H x 1-2'W.

BLUEBERRY FLAX
(Dianella tasmanica 'variegata')

Variegated green and white striped two-inch wide foliage. Sun/part shade. Tiny blue flowers in spring turn to blueberry-looking fruit. Turns brown in a freeze. Full sun/part shade. 3'H x 3'W.

LORIOPE
(Liriope muscari)

A Liriope for part sun or shade. Green foliage with purple flowers in summer. 'Big Blue' is 1-2'H x 1-2'W, 'Super Blue' is 2-3'H x 2-3'W. 'Variegated Liriope' has pretty yellow and green foliage and is 1-2'H x 1-2'W.

MUHLY GRASS
(Muhlenbergia capillaris)

Gorgeous soft pink plumes in the fall. A white variety also exists. Tolerates wet and dry conditions. Prune back to 6-12" in spring if desired. 3'H x 3'W. [N]

Part 3 - Plants

PAMPAS GRASS
(Cortaderia selloana)

Pretty with beautiful large soft white plumes. Beach-like. Full sun. Very tall and wide, rarely used in landscapes due to its size. 8-10'H x 5-8'W.

RED FOUNTAIN GRASS
(Pennisetum setaceum 'Rubrum')

Dark red grass with reddish-white plumes in late summer to fall. 2-4'H x 2-3'W.

GROUND COVERS

The term "ground cover" can be used in different ways. Some references consider tall grasses and other plants as ground covers. But here we are referring to the more common definition of ground covers: low-growing plants meant for filling in and covering an area. They are mostly used in community rather than for the front yard landscape. You can find low-growing landscape plants in the Shrub, Perennial, and Grass sections.

JASMINE, MINIMA
(Trachelosperum asiaticum)

Minima

The species Minima is a low-growing jasmine with green foliage and small, pale yellow, star-shaped, very fragrant flowers that rarely bloom. 'Summer Sunset' is a variety that has nice red, orange, yellow, and green foliage, and requires full sun for best color, although it also does well in part shade. Keep both trimmed for a smaller plant if desired. 20'H x 1½'W.

'Summer Sunset'

OTHERS

Frogfruit and Twin Flower, native plants with blue flowers, are mostly used as native lawn or mulch replacements. Sunshine Mimosa, with pink powderpuff flowers, is also native but will take over aggressively. Perennial Peanut has yellow flowers from summer through fall. It, too, spreads, so is mostly found in roadway medians, where it is automatically contained. It can be mowed. Even as a grass replacement, you will find weeds within these groundcovers.

'Perennial Peanut'

VINES

Vines are good for covering trellises, a pretty look which also provides privacy. Although people sometimes use them to climb a lamppost, most of these vines get tall and can block out the light and any lamppost signs, unless regularly pruned. Featured below are those commonly used in landscapes.

ALLAMANDA VINE
(Allamanda cathartica)

Beautiful showy yellow trumpet flowers in summer-fall. Full sun/part shade.

BLEEDING HEART
(Clerodendrum thomsoniae)

A part shade vine with clusters of red flowers with puffy white bracts, spring through fall. Moist soil.

BOUGAINVILLEA
(Bougainvillea spp.)

Bush

Many bright or pastel-colored bracts in pink and other colors. Blooms in winter and sporadically throughout the year. Full sun and dry conditions are needed for bloom. Most varieties have large spines – be careful of placement and while pruning (in spring). Thornless varieties can be found, as well as dwarf/bush varieties. Use on a strong trellis or keep trimmed as a shrub. If chewed branch tips or loss of leaves, check for the Bougainvillea looper caterpillar.

Vine

CAPE HONEYSUCKLE
(Tecomaria capensis)

Ann's Cape Honeysuckle

Not a true honeysuckle. Orange (or less common yellow) tubular flowers in spring and sporadically through summer attract hummingbirds. This vine is aggressive and is best planted in a container. Frequently sold as a standard. Full sun. [H]

CAROLINA JESSAMINE
(Gelsemium sempervirens)

Also known as Yellow Jassmine. A native vine with scented yellow trumpet flowers and glossy evergreen foliage. Blooms late winter to spring. Train to a trellis. Sap is an irritant and plant is toxic to pets, use gloves when pruning. Full sun. [H, N]

CORAL HONEYSUCKLE
(Lonicera sempervirens)

Tubular red-orange flowers that hummingbirds love. Takes time to become established. [BU, H, N]

JASMINE, CONFEDERATE
(Trachelospermum jasminoides)

Extremely fragrant (especially in the evening) white flowers in spring. Can perfume an entire yard. Fast growing, will cover a large area in no time, great for a privacy pergola. Full sun/part shade.

Year 1 — *Year 3*

MANDEVILLA
(Mandevilla splendens)

Usually pink tubular flowers, vining habit. Many like to grow this up their lamppost. Milky sap can irritate. Full sun. A zone 10 plant that usually succumbs to a freeze.

PASSION VINE
(Passiflora incarnata)

Attracts hummingbirds and is a host plant for Zebra Longwing and Gulf Frittary butterflies. Purple flowers with white and maroon centers from summer through fall. Will spread. [BU, N]

SHADE FOLIAGE

Although smaller ferns are typically used as houseplants, some ferns can be grown outdoors in Central Florida, mostly in part or full shade. Ferns have spores on the undersides of their leaves that are sometimes mistaken as pests.

There are other plants that are commonly used in shade as well. Several of the most common ferns and plants for shade landscaping are listed below.

AMARYLLIS
(Hippeastrum x hybridum)

Amaryllis bulbs have large strapping leaves and beautiful very large lily flowers, many in striking red colors. They are forced into bloom at the holidays, but most varieties bloom naturally in the spring. After you have enjoyed your holiday blooms indoors, plant outdoors in a shady area for future blooms. Plant so the top half of the bulb is above the soil.

BROMELIAD
(Bromeliaceae spp.)

Many varieties with colorful variegated foliage and striking upright flowers in red, pink, and yellow. After the flower dies, the plant will produce pups. Afternoon shade.

CALADIUM
(Caladium x hortulanum)

Caladiums are known for their colorful foliage, with various mixes of white, green, pinks, and reds. Foliage will go dormant underground in the winter months. 2'H x 2'W.

GROUND ORCHID
(Spathoglottis plicata)

Purple wispy flowers above strapping foliage spring through fall. 18"H x 12"W.

HOSTA
(Hosta 'SunHosta')

'Sun Hosta' is a medium sized hosta bred for hot, humid southeast conditions. However, you may find it does best in part sun in Central Florida. Green leaves have variegated white trim, and white fragrant flowers top stalks in early summer. Just like in northern parts of the country, they will go dormant over the winter in Central Florida. 2'H x 2'W.

KIMBERLY QUEEN FERN
(Nephrolepis obliterate)

Although this is a popular landscape fern because it can be grown in sun as well as shade, beware – it spreads aggressively and will take over a space in no time. It is best contained in a pot. 2'H x 3'W (or more).

PEACOCK GINGER
(Kaempferia pulchra/rotunda)

The southern version of a hosta, with green leaves that have purplish-bronze or silver markings and purple undersides. Small pink, purple, or white flowers appear in summer. They go dormant underground in winter. 1-2'H x 1-2'W.

Part 3 - Plants

RESURRECTION FERN
(Pleopeltis polypodioides)

Not normally grown in landscapes, rather in wooded areas on tree limbs, but it is worth mentioning here because it is a cool plant. When dry, it looks brown, shriveled up, and dead. But after rainfall, it comes alive again and turns green. [N]

EDIBLES

There are many edible plants that will grow in Central Florida. Some we cannot grow here as we could up north because of chilling requirements. Others we can grow here year-round because of our temperate climate. As a space-saving and cool-looking feature, sometimes fruits are grafted together and grown as a single plant, called "fruit cocktail trees." Citrus, in particular, have extremely fragrant flowers.

Edibles have their own special care requirements, and details are outside the scope of this book. Below are some of the most common plants for our area.

FRUIT COCKTAIL TREES

Citrus (lemon, lime, orange)

Fruit (apples, pears)

Stone fruit (peaches, plums, nectarines)

CITRUS TREES

Lemon, Meyer

Lemon, Ponderosa

Lime

Orange - many varieties

Part 3 - Plants

FRUITS TREES

Avocado, Florida (firmer and less fat than Hass)
Fig
Olive
Peach
Persimmon
Pomegranate

FRUITS

Bananas
Grapes, Muscadine
Pineapple

BERRIES

Blueberry
Blackberries
Strawberries

VEGETABLES

See my other book, Growing Fresh Vegetables.

HERBS

See my other book, Growing Fresh Herbs.

INDEX OF PLANTS

Abelia 151
African Iris 182
Agapanthus 183
Agave 209
Allamanda 164
Allamanda Vine 218
Allysum 200
Amaryllis 223
Angelonia 183
Anise 151
Arizona Cypress 137
Azalea 152
Aztec Grass 212
Bald Cypress 137
Bamboo 138
Beautyberry 153
Begonia 203
Birch, River 138
Bird of Paradise 165
Bismark Palm 127
Blackberry Lily 184
Blanket Flower 184
Bleeding Heart 218
Blueberry Flax 213
Blue Daze / Blue My Mind 185
Bottlebrush, Dwarf 153
Bottlebrush Tree 139
Bougainvillea 219

Part 3 - Plants

Bromeliad 224
Bulbine 185
Bush Daisy 165
Buttercup Bush 166
Butterfly Bush, Dwarf 154
Caladium 224
Camellia 139
Camellia, dwarf 154
Canary Island Date Palm 128
Canna Lily 186
Cape Honeysuckle 220
Cardboard Palm 136
Carolina Jessamine 220
Cassia 140
Chaste 140
Chinese Fan Palm 128
Chrysanthemum 187
Cleyera 141
Coleus 204
Coneflower 187
Coontie Palm 136
Copper Leaf 166
Coral Honeysuckle 221
Coreopsis 188
Crepe Myrtle 141
Crocosmia 188
Crossandra 189
Croton 167
Crown of Thorns 168
Dahlia 189

Daylily	190
Delphinium	200
Dewdrop	169
Dianthus	191
Dipladenia	169
Dwarf Palmetto	129
Esperanza	170
European Fan Palm	129
False Bird of Paradise	170
Firebush	171
Firecracker Fern	172
Foxtail Fern	155
Foxtail Palm	130
Gardenia	155
Gaura	191
Geranium	200
Gerbera Daisy	192
Ginger, Variegated/Shell	173
Ground Orchid	225
Hibiscus	173
Holly, Buford Dwarf	156
Holly, Shillings	156
Holly Tree	142
Hong Kong Orchid Tree	143
Hosta	225
Hydrangea, Oakleaf	157
Impatiens	204
Indian Hawthorne	157
Italian Cypress	144
Ixora	174

Part 3 - Plants

Japanese Blueberry	143
Jasmine, Confederate	221
Jasmine, Minima	216
Jatropha	175
Juniper	158
Kimberly Queen Fern	226
Lady Palm	130
Lantana	192
Ligustrum, Dwarf	158
Ligustrum Tree	144
Livingstone Daisy	193
Lobelia	205
Loriope	213
Loropetalum	159
Lysmachia	193
Magnolia	145
Mandevilla	222
Maple, Florida Red	145
Marigold	205
Mexican Heather	175
Mexican Petunia	159
Milkweed	194
Million Bells	201
Mona Lavender	194
Muhly Grass	214
Mussaenda	176
Nandina	160
Nemesia	201
Norfolk Island Pine	146
Oak	146

Oleander 160
Oyster Plant 176
Palms for Indoors or Lanais 135
Pampas grass 215
Pansy 201
Passion Vine 222
Peacock Ginger 226
Penta 195
Petunia 202
Philodendron 177
Pindo Palm 131
Pine 147
Pittosporum 161
Plumbago 177
Podocarpus 147
Podocarpus, Dwarf 161
Portulaca 206
Powderpuff 148
Prickly Pear 209
Purple Leaf Plum 148
Queen Palm 131
Red Fountain grass 215
Resurrection Fern 227
Ribbon Palm 132
Roebellini Palm 132
Rose 162
Rudbeckia 195
Sabal Palm 133
Sago Palm 136
Salvia 196

Part 3 - Plants

Saw Palmetto	133
Scaevola	206
Sedum	210
Serissa	178
Shefflera/Trinette	178
Shrimp Plant	179
Snapdragon	202
Society Garlic	196
Stoke's Aster	197
Succulents, Tropical	210
Sylvester Palm	134
Tea Olive	149
Texas Sage	163
Thyrallis	179
Tibouchina	180
Ti Plant	180
Trumpet Tree	149
Verbena	197
Viburnum	150
Vinca	198
Walter's Viburnum	163
Windmill	134
Yesterday, Today, Tomorrow	181
Yucca, Dwarf	211
Yucca, Spineless	211
Zinnia	207

EPILOGUE

I hope you enjoyed our journey through the wonderful world of landscaping and plants in Central Florida. Happy gardening!

ACKNOWLEDGMENTS

Thank you to my wonderful friends, including Alycyn, Ann, Anne, Cat, Connie, Diane, Ken, Marti, Nan, Nini, and Tanya, who allowed me to take or use photos of their beautiful plants/yards to use in this book. I am lucky to be in a community where so many homeowners desire a beautiful Central Floridian yard, which in turn provides bountiful photo opportunities. Thanks to Sabrina for sharing her knowledge and answering my technical questions over the years. Thanks to Ray and Wayne for helping me to learn my southern plants back in the beginning. Thanks to my publisher, Nancy, editor, Vickie, and formatter, Amber, who all ensured that my books have that professional touch.

Much effort was extended from many hours of writing, and countless miles were logged driving around to take photos. So lastly, thanks to my husband for understanding the continual "I can't, I'm working". With the completion of this, my third gardening book in the series, it is now back to a normal life.

ABOUT THE AUTHOR

Rondi is a horticulturist and Florida master gardener. For two decades she worked in the horticultural field, including at various local garden centers as a grower and seller of plants, as a grower of organic herb and vegetable plants in her own greenhouses for farmers' markets, and by providing on-site consultations for landscape design and plant diagnostics.

As a master gardener coordinator, she taught future enthusiasts the knack of gardening. And she has been teaching gardening locally for over a decade. She is also the chairperson of the Landscape & Garden Club for her community.

She resides in The Villages, Florida, with her husband and menagerie of pets, surrounded by her tropical oasis of plants, gardens, and natural woods.

Made in the USA
Columbia, SC
08 February 2025